Lecture Notes in Mathematics 1725

Editors:
A. Dold, Heidelberg
F. Takens, Groningen
B. Teissier, Paris

Springer
Berlin
Heidelberg
New York
Barcelona
Hong Kong
London
Milan
Paris
Singapore
Tokyo

Dieter A. Wolf-Gladrow

Lattice-Gas Cellular Automata and Lattice Boltzmann Models

An Introduction

Springer

Author

Dieter A. Wolf-Gladrow
Alfred Wegener Institute for Polar and Marine Research
Postfach 12 01 61
27515 Bremerhaven, Germany
E-mail: dwolf@awi-bremerhaven.de

Cataloging-in-Publication Data applied for

Die Deutsche Bibliothek - CIP-Einheitsaufnahme

Wolf-Gladrow, Dieter:
Lattice gas cellular automata and lattice Boltzmann models : an
introduction / Dieter A. Wolf-Gladrow. - Berlin ; Heidelberg ; New
York ; Barcelona ; Hong Kong ; London ; Milan ; Paris ; Singapore ;
Tokyo : Springer, 2000
(Lecture notes in mathematics ; 1725)
ISBN 3-540-66973-6

Mathematics Subject Classification (1991): 35Q30, 58F08, 65C20, 76P05, 82C40

ISSN 0075-8434
ISBN 3-540-66973-6 Springer-Verlag Berlin Heidelberg New York

Springer-Verlag Berlin Heidelberg New York
a part of SpringerScience+Business Media

© Springer-Verlag Berlin Heidelberg 2000
Printed in Germany

Typesetting: Camera-ready TeX output by the author
Printed on acid-free paper   41/3111-5432

Table of Contents

1. Introduction

1.1 Preface

Lattice-gas cellular automata (LGCA)[1] and even more lattice Boltzmann models (LBM) are relatively new and promising methods for the numerical solution of (nonlinear) partial differential equations. Each month several papers appear with new models, investigations of known models or methodically interesting applications. The field of lattice-gas cellular automata started almost out of the blue in 1986 with the by now famous paper of Frisch, Hasslacher and Pomeau. These authors showed, that a kind of billiard game[2] with collisions that conserve mass and momentum, in the macroscopic limit leads to the Navier-Stokes equation when the underlying lattice possesses a sufficient (hexagonal in two dimensions) symmetry. A few years later lattice Boltzmann models arose as an offspring of LGCA. Their higher flexibility compared to LGCA led to artificial microscopic models for several nonlinear partial differential equations including the Navier-Stokes equation.

I have followed the exciting development of both methods since 1989 and from time to time have given courses on this topic at the Department of Physics and Electrical Engineering at the University of Bremen (Germany). The present book is an extended version of my lecture manuscript.

The word 'introduction' in the title implies two things. Firstly, the level of presentation should be appropriate for undergraduate students. Thus methods like the Chapman-Enskog expansion or the maximum entropy principle which are usually not taught in standard courses in physics or mathematics are discussed in some detail. Secondly, in an introduction many things have to be left out. This concerns, for instance, models with several colors which allow the simulation of multiphase flows[3] or magnetohydrodynamics. Only a few applications of LGCA or LBM to physical problems can be considered. Interesting topics like the divergence of transport coefficients in 2D are not discussed. The interested reader will find, however, references pointing to original articles (especially in the 'What else?' sections).

The lattice-gas cellular automata require special programming techniques which are only sparsely discussed in the widely scattered literature. The book will hopefully fill a gap in this respect (see Subsections 3.1.2, 3.1.4, 3.2.2). Several program codes will be made available via internet (http://www.awi-bremerhaven.de/Modelling/LGCA+LBM/index.html).

Many mathematical text books and courses contain lots of definitions, theorems and proofs - and not much else. In this respect the current book is rather 'unplugged': the emphasis is more focused on presenting the main principles

---

[1] The abbreviations are explained in Table 6.6.5 on page 270.

[2] Goldenfeld and Kadanoff (1999) compare the time development of lattice-gas cellular automata with a square dance.

[3] When I had almost finished my manuscript I became aware of the wonderful book by Rothman and Zaleski (1997). Simulation of multiphase flows is a major topic in that book.

and not on teaching proof techniques. Nonetheless the proofs of several essential theorems are presented in detail.

Last but not least, I would like to add a few comments on the exercises. Problems with one star (∗) should be very easy to solve (in a few minutes); those with two stars require more thinking or somewhat lengthy ('... after some algebra ...') calculations. Exercises with three stars are very different. Some of them require quite a bit of programming; others address more advanced stuff which has not been treated here. And finally, some of the three star exercises point to open problems which I have not solved myself.

Acknowledgements:

The following people supported me in one or the other way by teaching me mathematics and physics, introducing me to LGCA, providing PhD positions, asking stimulating questions, proofreading etc. Ernst Augstein, Uwe Dobrindt, Fritz Dröge, Lars-Peer Finke, Silvia Gladrow, Vladimir Gryanik, Wolfgang Hiller, Matthias Hofmann, Heiko Jansen, Charilaos Kougias, Gerrit Lohmann, Ferial Louanchi, Christof Lüpkes, Ralf Nasilowski, Dirk Olbers, Christoph Völker, Armin Vogeler, Werner Wrede, Richard Zeebe.
I am grateful to them all.
Three anonymous referees made useful comments.
I also thank Stefanie Zöller (Springer Verlag) for support.

1.2 Overview

The plan of the book is as follows (compage Fig. 1.2.1). In an introductory section the Navier-Stokes equation and several approaches to solve it are discussed. In Chapter 2 cellular automata (CA) are treated in some detail in order to show the special character of lattice-gas cellular automata. CA rules are usually not restricted by conservation laws which is a nice feature when simulating growth processes. The spatial propagation of properties is part of the local updating rule. In contrast, lattice-gas cellular automata obey certain conservation laws and the updating is splitted into a local 'collision' and a propagation to the nearest-neighbor sites. This splitting makes it easier to construct models with desired macroscopic properties. The CA chapter can be skipped in first reading.

Chapter 3 on lattice-gas cellular automata starts with the historically first LGCA, namely the HPP model. This is the simplest model that aimed to simulate the Navier-Stokes equation (but failed to do so!). The emphasis here is on a discussion without digging too much into theory. Special programming techniques like multi-spin coding are explained in detail.

The FHP model is the first successful LGCA. Starting from the Boolean microdynamics the macroscopic equations will be derived up to first order (Euler equation) by a multi-scale expansion (Chapman-Enskog). The second order which yields the Navier-Stokes equation will be addressed later on in the chapters on statistical mechanics (Section 4.2) as well as in the one on lattice Boltzmann models (Section 5.2.3).

The difference between failure (HPP) and success (FHP) depends on the symmetry of the underlying lattice. The tensor of rank four formed from products of the lattice vectors is part of the advection term and has to be isotropic. The main problem in proposing a LGCA for simulations of flows in three dimensions is to find a lattice with sufficient symmetry. In Section 3.3 the lattice tensors of rank two and four for several lattices will be calculated and investigated for isotropy.

If one restricts oneself to single-speed models the only lattice feasible for three-dimensional simulations is the four-dimensional face-centered hyper-cube (FCHC). Several possible collision rules for this model are outlined in Section 3.5. As an alternative to FCHC multi-speed models are available. When the collision rules are carefully chosen these models conserve energy in addition to mass and momentum and therefore are called thermal models (Section 3.7). Another alternative for simulation in 3D is the pair-interaction (PI) model (Section 3.6). The collision rules of this model are simple in 2D as well as in 3D and thus allow coding using bit-operators.

In Chapter 4 some relevant concepts from statistical mechanics are discussed. Specifically the Boltzmann equation, its five collision invariants, and its (global) equilibrium distribution (Maxwell-Boltzmann) are presented. The chapter contains a proof of Boltzmann's famous H-theorem. For many applications the complicated collision integral can be substituted by a relaxation toward equilibrium by a term that is proportional to the deviation of the actual distribution from its (local) equilibrium. With this so-called BGK approximation it is possible to derive the Navier-Stokes equation by the Chapman-Enskog expansion on few pages (Section 4.2). In addition, this chapter contains a section on the maximum entropy principle which will be applied later on in the derivation of equilibrium distributions for lattice Boltzmann models.

Chapter 5 is devoted to lattice Boltzmann models. This chapter is almost selfcontained. Readers who are only interested in LBMs (and not in LGCA) can start here but should read at some point Section 3.3 on lattice tensors. However, some remarks in this chapter only make sense to those who are familiar with LGCA.

In Section 5.1 some problems with LGCA are listed and the transition from LGCA to LBM is sketched. The section on the D2Q9 model is in some respect the pendant to the FHP[4] section in Chapter 3 in that this BGK model is discussed in full detail. The equilibrium distributions are calculated from the maximum entropy principle, the Navier-Stokes equation is derived by Chapman-Enskog expansion and implementations of various boundary conditions are discussed. This model is applied to ocean circulation in Section 5.7. The stability of the D2Q9 and other LBMs is discussed in Section 5.6.

Although the use of the maximum entropy principle is a very elegant method, it hides the much wider freedom in choosing equilibrium distributions. Alternatively, one may start from a reasonable ansatz for the distributions and then fix the free parameters during or after the multi-scale expansion such that the desired equations (Navier-Stokes or other partial differential equations) are obtained (Section 5.4).

This ansatz method is used to derive LBMs for diffusion equations (linear as well as nonlinear in any number of dimensions) in Section 5.8. These models can easily be extended to LBMs for reaction-diffusion equations. With the same methods thermal LBMs can be constructed (Section 5.5). LBMs for simulation in 3D are described in Section 5.3. The appendix contains a section on Boolean algebra, some lengthy calculations and code listings of FHP collision rules.

---

[4] Although the underlying lattices are different!

Fig. 1.2.1. *Overview*

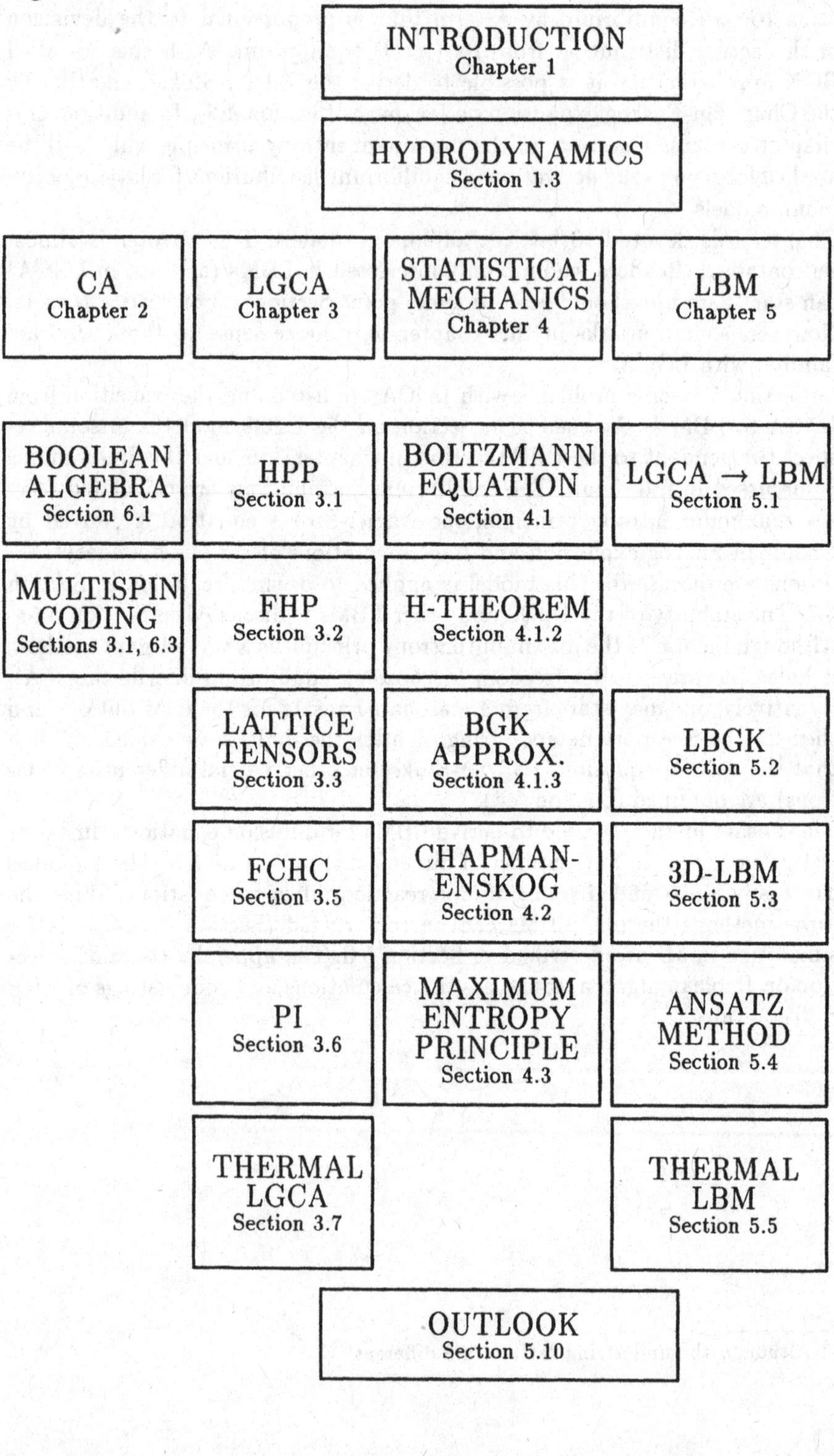

	INTRODUCTION Chapter 1		
	HYDRODYNAMICS Section 1.3		
CA Chapter 2	LGCA Chapter 3	STATISTICAL MECHANICS Chapter 4	LBM Chapter 5
BOOLEAN ALGEBRA Section 6.1	HPP Section 3.1	BOLTZMANN EQUATION Section 4.1	LGCA → LBM Section 5.1
MULTISPIN CODING Sections 3.1, 6.3	FHP Section 3.2	H-THEOREM Section 4.1.2	
	LATTICE TENSORS Section 3.3	BGK APPROX. Section 4.1.3	LBGK Section 5.2
	FCHC Section 3.5	CHAPMAN-ENSKOG Section 4.2	3D-LBM Section 5.3
	PI Section 3.6	MAXIMUM ENTROPY PRINCIPLE Section 4.3	ANSATZ METHOD Section 5.4
	THERMAL LGCA Section 3.7		THERMAL LBM Section 5.5
	OUTLOOK Section 5.10		

1.3 The basic idea of lattice-gas cellular automata and lattice Boltzmann models

Lattice-gas cellular automata (LGCA) and lattice Boltzmann models (LBMs) are methods for the simulation of fluid flows[5] which are quite distinctive from molecular dynamics (MD) on the one hand and methods based on the discretization of partial differential equations (finite differences, finite volumes, finite elements, spectral methods) on the other hand. Here the basic idea of LGCA and LBM will be sketched and the differences compared to other methods will be outlined.

1.3.1 The Navier-Stokes equation

The flow of incompressible fluids can be described by the Navier-Stokes equation[6]

$$\frac{\partial u}{\partial t} + (u\nabla)u = -\nabla P + \nu\nabla^2 u \tag{1.3.1}$$

together with the continuity equation[7]

$$\nabla \cdot u = 0 \tag{1.3.2}$$

where ∇ is the nabla operator, u is the flow velocity, $P = p/\rho_0$ the kinematic pressure, p the pressure, ρ_0 the constant mass density and ν the kinematic shear viscosity. Different fluids like air, water or olive oil are characterized by their specific values of mass density and viscosity ($\nu_{air} = 1.5 \cdot 10^{-5}$ m^2 s^{-1}, $\nu_{water} = 10^{-6}$ m^2 s^{-1}, $\nu_{olive\ oil} = 10^{-4}$ m^2 s^{-1}). Incompressible flows of these fluids obey the same form of equation (Navier-Stokes) whereas their microscopic interactions are quite different (compare gases and liquids!). The Navier-Stokes equation is nonlinear in the velocity u which prohibits its analytical solution except for a few cases. Numerical methods are required to

---

[5] ... and several other processes which can be described on the macroscopic level by partial differential equations ...

[6] The viscous term of the equation was derived in different ways by Claude Louis M. H. Navier (1785-1836) and Sir George Gabriel Stokes (1819-1903). The Navier-Stokes equation in tensor notation reads:

$$\partial_t u_\alpha + u_\beta \partial_{x_\beta} u_\alpha = -\partial_{x_\alpha} P + \nu \partial_{x_\beta} \partial_{x_\beta} u_\alpha.$$

[7] $\nabla \cdot u = 0$ is derived from the general continuity equation

$$\frac{\partial \rho}{\partial t} + \nabla \cdot (\rho u) = 0$$

by setting $\rho = $ constant.

simulate the time evolution of flows. On the other hand, the nonlinear advection term is most welcome because it is responsible for many interesting phenomena such as solitons (nonlinear waves), von Karman vortex streets (regular vortex shedding behind an obstacle) or turbulence.

The Reynolds number and dynamic similarity of flows. Flows with small velocities are smooth and are called laminar. At very high velocities they become turbulent. The transition from laminar to turbulent flows does not depend only on velocity as will be shown below. Consider the flow past an obstacle, such as a sphere, a cylinder or a plate. What are the characteristic scales of the flow? Obviously the flow field will depend on the (unperturbed) upstream speed U and the linear size (diameter L) of the obstacle. The fluid is characterized by its kinematic viscosity ν. The three parameters U, L and ν have dimensions [length time^{-1}], [length] and [length^2 time^{-1}]. It is easy to see that from these parameters one can form essentially one dimensionless number, namely the Reynolds number

$$R_e = \frac{UL}{\nu}. \tag{1.3.3}$$

The parameters U and L can be used to scale all quantities in the Navier-Stokes equation (the primed quantities are measured in units of U and L): $u' = u/U$, $x' = x/L$, $\nabla' = L \cdot \nabla$, $\nabla'^2 = L^2 \cdot \nabla^2$, $t' = t \cdot U/L$ (the advection time scale L/U is the time for the unperturbed flow to pass the linear size of the obstacle), $P' = P/U^2$ (the kinematic pressure has the dimension of energy per mass). Inserting the scaled quanties into the Navier-Stokes equation leads to

$$\frac{\partial u'}{\partial t'}\frac{U^2}{L} + (u'\nabla')\,u'\frac{U^2}{L} = -\nabla'P'\frac{U^2}{L} + \nu\nabla'^2 u'\frac{U}{L^2}$$

or after division by U^2/L

$$\frac{\partial u'}{\partial t'} + (u'\nabla')\,u' = -\nabla'P' + \frac{1}{R_e}\nabla'^2 u'. \tag{1.3.4}$$

The scaled Navier-Stokes equation equation (1.3.4) does not contain any scale and only one dimensionless quantity, namely the Reynolds number. Thus for a given type of flow (say the flow past a sphere) the scaled velocity of a stationary flow will depend only on the scaled spatial coordinate and the Reynolds number:

$$u' = \frac{u}{U} = f_u\left(\frac{x}{L}, R_e\right) \tag{1.3.5}$$

where the function f_u depends on the geometry of the problem (the type of flow). The same is true for the scaled pressure:

$$P' = \frac{P}{U^2} = f_P\left(\frac{x}{L}, R_e\right). \tag{1.3.6}$$

Thus all flows of the same type but with different values of U, L and ν are described by one and the same non-dimensional solution (u', P') if their Reynolds numbers are equal. All such flows are said to be *dynamically similar*.

The value of the Reynolds number provides an estimate of the relative importance of the non-viscous and viscous forces. The pressure gradient usually plays a passive role, being set up in the flow as a consequence of motions of a rigid boundary or of the existence of frictional stresses (Batchelor, 1967). Thus the flows can be characterized by the relative magnitudes of advection and viscous forces:

$$\frac{|(u\nabla)\,u|}{|\nu\nabla^2 u|} \approx \frac{U^2/L}{\nu U/L^2} = \frac{U \cdot L}{\nu} = R_e. \qquad (1.3.7)$$

Flows with small Reynolds numbers ($R_e \ll 1$) are laminar, von Karman vortex streets are observed at intermediate values ($R_e \approx 100$) and turbulent flows occur at very high Reynold numbers ($R_e \gg 1$). The fact that flows can be characterized by R_e and the *law of dynamic similarity* were first recognized by Stokes (1851) and Reynolds (1883).

The law of dynamic similarity provides the link between flows in the real world where length is measured in meters and the simulation of these flows with lattice-gas cellular automata and lattice Boltzmann models over a lattice with unit grid length and unit lattice speed. In these models the viscosity is a dimensionless quantity because it is expressed in units of grid length and lattice speed. These dimensionless flows on the lattice are similar to real flows when their Reynolds numbers are equal.

1.3.2 The basic idea

The fact that different microscopic interactions can lead to the same form of macroscopic equations is the starting point for the development of LGCA. In addition to real gases or real liquids one may consider artificial micro-worlds of particles 'living' on lattices with interactions that conserve mass and momentum. The microdynamics of such *artificial micro-worlds* should be very simple in order to run it efficiently on a computer. Consider, for example, a square lattice with four cells at each node such that one cell is associated with each link to the next neighbor node (compare Fig. 3.1.1 on page 41). These cells may be empty or occupied by at most one particle with unit mass $m = 1$. Thus each cell has only two possible states and therefore is called a *cellular automaton*. Velocity and thereby also momentum can be assigned to each particle by the vector connecting the node to its next neighbor node along the link where the particle is located. These vectors are called lattice velocities. The microscopic interaction is strictly local in that it involves only particles at a single node. The particles exchange momentum while conserving the mass and momentum summed up over each node. After this *collision* each particle propagates along its associated link to its next neighbor node. The microdynamics consists on a repetition of collision and propagation. Macroscopic values of mass and momentum density are calculated by *coarse graining* (calculation of mean values over large spatial regions with hundreds to thousands of nodes).

Do these mean values obey the Navier-Stokes equation? The answer is negative for the model just sketched (discussed in more detail in Section 3.1). This model was proposed by Hardy, de Pazzis and Pomeau in 1973 (HPP model). It took more than 10 years before Frisch, Hasslacher and Pomeau (1986) found the third essential condition in addition to mass and momentum conservation: the lattice has to possess a sufficient *symmetry* in order to ensure *isotropy* of a certain tensor of fourth rank formed from the lattice velocities. In 2D, for example, 4-fold rotational symmetry (square lattice) is not enough whereas hexagonal symmetry (triangular lattice; FHP model; see Section 3.2) is sufficient.

A further condition should be mentioned here. The microdynamics must not possess more invariants than required by the desired macroscopic equations because such so-called *spurious invariants* can alter the macroscopic behavior by unphysical constraints (compare Section 3.8).

The importance of the work of Frisch, Hasslacher and Pomeau (1986) can hardly be overestimated. Their finding of the lattice symmetry condition started an avalanche of LGCA models. Finding a lattice with sufficient symmetry for simulations in 3D was a tough job. Wolfram (1986) showed that lattice tensors over the *face-centered hypercube* (FCHC) are isotropic up to rank 4.

Lattice Boltzmann models were first based on LGCA in that they used the same lattices and applied the same collisions. Instead of particles, LBMs deal with continuous distribution functions which interact locally (only distributions at a single node are involved) and which propagate after 'collision' to the next neighbor node. Coarse graining is not necessary any more. In the beginning this was considered as the main advantage of LBMs compared to LGCA. The next step in the development was the simplification of the collision operator and the choice of different distribution functions. This gives much more flexibility of LBMs, leads to Galilei invariant macroscopic equations without scaling of time, and allows to tune viscosity. Most recently LBMs living on curvilinear coordinate systems have been proposed.

Exercise 1.3.1. (**)
Consider flows that are affected by an external force such as gravity. Discuss the consequences for the similarity of flows. How many independent dimensionless numbers are required to characterize the flow?

1.3.3 Top-down versus bottom-up

The conventional simulation of fluid flows (and other physical processes) generally starts from nonlinear partial differential equations (PDEs). These PDEs are discretized by *finite differences* (Ames, 1977; Morton and Mayers, 1994), *finite volumes* (Bryan, 1969), *finite elements*[8] (Zienkiewicz, and Taylor, 1989 and 1991), or *spectral methods* (Machenhauer, 1979; Bourke, 1988). The resulting algebraic equations or systems of ordinary differential equations are solved by standard numerical methods. Although this 'top-down' approach seems to be straightforward it is not without difficulties. In many textbooks on the numerical solution of partial differential equations the authors put much emphasis on the truncation error which is due to the truncation of Taylor series when going from differential to finite differences whereas physicists are usually more concerned whether or not certain quantities are conserved also by the discretized form of the equations. This latter property is most important for integrations over long time scales in closed domains like, for instance, in the simulation of the world oceans or in coupled atmosphere-ocean models. A small leakage would transform the ocean into an empty basin after some time. *Numerical instabilities* are another problem of this type of numerical methods (Courant, Friedrichs and Lewy, 1928; Phillips, 1956 and 1959).

LGCA and LBM are different variants of the 'bottom-up' approach (Fig. 1.3.1) where the starting point is a discrete microscopic model which by construction conserves the desired quantities (mass and momentum for Navier-Stokes equation). These models are unconditional stable (LGCA) or show good stability properties (LBM). The derivation of the corresponding macroscopic equations requires, however, lengthy calculations (multi-scale analysis). A major problem with the bottom-up approach is to detect and avoid spurious invariants which is, by the way, also a problem for the models derived by the top-down approach. The construction of LGCA or LBM for given macroscopic equations seems to require some intuition. Meanwhile at least for LBM there exists a recipe for the construction of appropriate microdynamics when the conserved quantities of the physical process are known (compare Section 5.4).

1.3.4 LGCA versus molecular dynamics

Another bottom-up approach is *molecular dynamics* (MD) (Verlet, 1967; Evans and Morriss, 1983; Heyes et al., 1985; Mareschal and Kestemont, 1987; Boon and Yip, 1991; Rapaport, 1995). In MD one tries to simulate macroscopic behavior of real fluids by setting up a model which describes the microscopic interactions as good as possible. This leads to realistic equations of

---

[8] These finite methods can be combined with multigrid techniques; see, for example, Hackbusch (1985).

state whereas LGCA or LBM posses only isothermal relations between mass density and pressure. The complexity of the interactions in MD restricts the number of particles and the time of integration. A method somewhat in between MD and LGCA is *maximally discretized molecular dynamics* proposed by Colvin, Ladd and Alder (1988).

Fig. 1.3.1. *Top-down versus bottom-up (see text).*

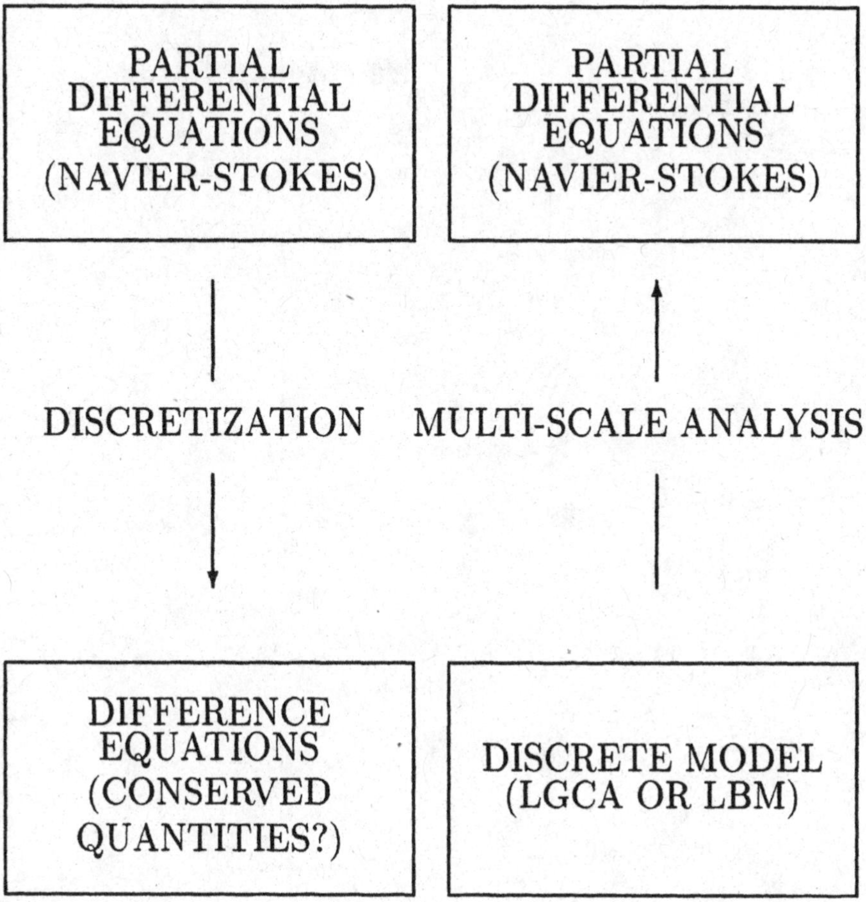

2. Cellular Automata

> "Cellular automata are sufficient simple to allow detailed mathematical analysis, yet sufficient complex to exhibit a wide variety of complicated phenomena."
> Wolfram (1983)

Cellular automata (CA) cannot and should not be covered comprehensively in this book. The current chapter shall give the reader a glimpse of the manifold arrangements and the peculiarities of CA. It will serve as a background for the discussion of a special type of cellular automata, namely lattice-gas cellular automata.

2.1 What are cellular automata?

CA can be characterized as follows (e.g., Wolfram, 1984b or Hedrich, 1990; see below for a formal definition):

- CA are regular arrangements of single *cells* of the same kind.
- Each cell holds a finite number of discrete states.
- The states are updated simultaneously ('synchronously') at discrete time levels.
- The update rules are deterministic and uniform in space and time.
- The rules for the evolution of a cell depend only on a local neighborhood of cells around it.

Not all of these criteria are always fulfilled. The cells can be positioned, for example, at the nodes of a (quasiperiodic) *Penrose lattice* (Penrose, 1974, 1979) or at random (Markus and Hess, 1990). A random connection of cells was proposed by Richard Feynman (Hillis, 1989). The update rules of certain CA include probabilistic elements (compare the FHP lattice gas automata, Section 3.2).

The formal definition of CA follows Kutrib et al. (1997). The cells can be imagined as positioned at the integer points of the D-dimensional Euclidean *lattice* $\mathcal{L} = \mathbb{Z}^D$. The *finite set of possible states* of each of the cells is equal and will be denoted by Q[1].

The state of a cell i at a new time level $t + 1$ depends on the states of cells j in a finite neighborhood[2] $N \subset \mathbb{Z}^D$ at time t[3]. The elements $n \in N$ are to be interpreted as the *relative coordinates* of neighboring cells (with $(0, ..., 0)$ as relative coordinate of cell i). The neighborhood N may be interpreted as an interconnection between the cells.

A mapping $l : N \rightarrow Q$ is called a *local configuration*[4]. It contains exactly the information to update a cell. The mode of operation of a cell is completely determined by its *local rule* $r : Q^N \rightarrow Q$ where Q^N is the set of all mappings $f : N \rightarrow Q$. The CA updating is called homogeneous when the neighborhoods N and N' of the cells i and i' map onto each other by a translation and when the same local rule is applied to all cells.

The *global configuration* of a CA (i.e. the ensemble of the state of all cells) at a certain time is called a (*global*) *configuration* g. CA are working in discrete time. The global configuration g at time t leads to a new global configuration g' at time $t + 1$ whereby all cells enter a new state according to the local rule *synchronously*. The associated *global rule* is a mapping $R : Q^{\mathcal{L}} \rightarrow Q^{\mathcal{L}}$.

2.2. A short history of cellular automata

Around 1950 cellular automata[5] were introduced by Stanislas Ulam, John von Neumann, and Konrad Zuse[6]. Ulam simulated the growth of patterns in two and three dimensions (compare Ulam, 1952 and 1962; Schrandt and Ulam, 1970). John von Neumann proposed a self-reproducing cellular automaton (von Neumann, 1966) which at the same time realized a universal *Turing machine* (Turing, 1936; Hopcroft, 1984). Each of the approximately 200000 cells of von Neumann's CA holds 29 different states. A few years ago this CA has been implemented for the first time on a computer (Signorini, 1989).

---

[1] For example, $Q = \{0, 1\}$ for a binary automaton

[2] The neighborhood includes the cell i. There are, however, particular update rules that do not depend on the state of i at time t.

[3] A random process whose future probabilities are determined by its most recent values is called a Markov process. If not otherwise stated the updating of CA is, however, a deterministic process.

[4] We will also denote an actual state of a neighborhood N as the local configuration.

[5] Other names: cellular spaces, tesselation automata, homogeneous structures, cellular structures, tesselation structures, iterative arrays.

[6] Some scientists even regard the paper by Wiener and Rosenblueth (1946) as the first one in this field.

Zuse published his ideas concerning the application of cellular automata to physical problems in a monograph (Zuse, 1969; English translation 1970). Some of his formulations already resemble to the HPP lattice-gas cellular automata proposed four years later by Hardy et al. (1973). In addition to hydrodynamic problems Zuse had in mind models for electrodynamics and quantum theory. The most far-reaching vision was his concept of the universe as a cellular automaton encompassing a gigantic number of cells (Zuse, 1982).

As far as number of citations can tell something about the flow of ideas, Zuse's monograph (1969; 1970) did not have a major impact (but see Alasyev et al., 1989; Case et al., 1990; Fredkin, 1990; Toffoli, T. and N. Margolus, 1990; Rothman and Zaleski, 1994).

In 1970 John Horton Conway introduced the game 'Life', a two-dimensional CA with simple update rules but complex dynamics (compare Section 2.4.3). Martin Gardner made cellular automata very popular by a series of papers on 'Life' in Scientific American (Gardner, 1970, 1971a,b,c; see also: Berlekamp, Conway and Guy, 1984).

The first lattice-gas cellular automata (LGCA) - special kinds of cellular automata for the simulation of fluid flow and other physical problems - was proposed in 1973 by Hardy, Pomeau and de Pazzis. Its name HPP is derived from the initials of the three authors. Although the HPP model conserves mass and momentum it does not yield the desired Navier-Stokes equation in the macroscopic limit.

In 1983 Stephen Wolfram revived the interest in CA by a series of papers (Wolfram, 1983, 1984a,b,c). The one-dimensional arrays of cells considered by Wolfram expressed complex patterns when initialized randomly and updated by simple deterministic rules depending on the state of the cell and a few of its neighbors.

In 1986 Frisch, Hasslacher and Pomeau discovered that a CA over a lattice with hexagonal symmetry, i.e. with a somewhat higher symmetry than for the HPP model, leads to the Navier-Stokes equation in the macroscopic limit. The theoretical foundations of lattice gas automata were given soon after by Wolfram (1986) and Frisch et al. (1987).

2.3 One-dimensional cellular automata

Wolfram (1983, 1984a,b) investigated one-dimensional cellular automata. He introduced a division into four universal classes. Even the study of the branch concerned with one-dimensional cellular automata is far from completed because only a small subset of possible rules has been explored and a theoretical understanding is still in its infancy (Wolfram, 1985).

One-dimensional cellular automata consist of a number of uniform cells arranged like beads on a string. If not stated otherwise arrays with finite number of cells and periodic boundary conditions will be investigated, i.e. the beads form a necklace (compare Fig. 2.3.1). The states of all cells form a (*global*) *configuration* of a CA.

Fig. 2.3.1. *One-dimensional cellular automata with the two possible states per cell: empty or occupied (marked with a cross).*

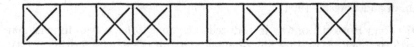

The state of cell i at time t is referred to as $a_i^{(t)}$. The finite number of possible states $k < \infty$ are labelled by non-negative integers from 0 to $k - 1$, i.e. $a_i^{(t)} \in \mathbb{Z}_k$ (mathematicians call the set of integers modulo k the residue class $\mathbb{Z}_k$). The state of each cell develops in time by iteration of the map

$$a_i^{(t)} = F[a_{i-r}^{(t-1)}, a_{i-r+1}^{(t-1)}, \ldots a_i^{(t-1)}, \ldots, a_{i+r}^{(t-1)}] \tag{2.3.1}$$

i.e. the state of the ith cell at the new time level t depends only on the state of the ith cell and the r (range) neighbors to the left and right at the previous time level $t - 1$. The arbitrary (in general nonlinear) function F is called the *automata rule*, the *update rule* or just the *rule*. An alternative formulation of the rule (2.3.1) reads

$$a_i^{(t)} = f\left[\sum_{j=-r}^{j=r} \alpha_j a_{i+j}^{(t-1)}\right] \tag{2.3.2}$$

where the α_j are integer constants and thus the function f has a single integer as argument.

Exercise 2.3.1. (**)
Why can (2.3.1) and (2.3.2) be equivalent formulations?

Number of automata rules. Consider a CA with $k = 2$ possible states per cell and a range $r = 1$. The possible combinations of the arguments of the automata rule F are listed in two different representations in Tables 2.3.1 and 2.3.2.

There are 8 different combinations (in general: k^{2r+1}). Interpretation of $\{a_{i-1}^{(t-1)}, a_i^{(t-1)}, a_{i+1}^{(t-1)}\}$ (columns 2 to 4 in Table 2.3.1) as the bit pattern (with the highest bit to the left) of an integer in binary representation yields the

Table 2.3.1. *An example of an automata rule for a CA with $k = 2$ and $r = 1$.*

	$a_{i-1}^{(t-1)}$	$a_i^{(t-1)}$	$a_{i+1}^{(t-1)}$	$a_i^{(t)}$
0	0	0	0	0
1	0	0	1	1
2	0	1	0	0
3	0	1	1	0
4	1	0	0	1
5	1	0	1	0
6	1	1	0	0
7	1	1	1	0

Table 2.3.2. *An example of an automata rule for a CA with $k = 2$ and $r = 1$.*

111	110	101	100	011	010	001	000
0	0	0	1	0	0	1	0

numbers 0 to 7 (listed in the first column). In the last column of Table 2.3.1 one of the possible rules is given in tabular form. It consists of a certain sequence of zeros and ones which also can be interpreted as the binary representation of an integer. Each bit pattern of length 8 corresponds to an automata rule. Therefore it follows immediately that there exist $2^8 = 256$ different rules (in general: $k^{k^{2r+1}}$). CA in 1D with updating rules depending only on the site itself and the sites immediately adjacent to it on the left and right will be denoted as *elementary cellular automata* (Wolfram, 1983, p. 603). Instead of the tabular form (bit pattern) the automata rules are often referred to by the corresponding integer between 0 and 255 which is called the *rule number*. Thus the rule 00010010 given in Table 2.3.1 is denoted as rule 18. Similar rule number can also be defined for automata with more than two states per cell. Because the number of rules rapidly increases with k and r (compare Table 2.3.3) only a small part of all possible rules has been investigated.

Subclasses of rules. Subclasses of rules can be obtained by applying the following definitions:

– *Additive* rules: f is a linear function of its argument modulo k. Remark: These rules obey a special additive superposition principle and therefore are accessible to an algebraic analysis (Martin et al., 1984).

Table 2.3.3. *The number of possible rules for cellular automata with k states per cell and a range r. Listed are only the cases where the number is smaller than 10^{100} (Gerling, 1990b).*

k/r	1	2	3
2	2^8	2^{32}	2^{128}
3	3^{27}	–	–
4	4^{64}	–	–
5	5^{125}	–	–

- *Totalistic* rules: $\alpha_j \equiv 1 \quad \forall j$ in (2.3.2), i.e. the cell and all its neighbors in the range r contribute equally and for $k = 2$ only the sum of occupied cells matters.

- *Symmetric* rules: $F[a_{i-r}, ..., a_{i+r}] = F[a_{i+r}, ..., a_{i-r}]$.

- Rules *with memory*: $a_i^{(t)}$ depends on $a_i^{(t-1)}$ (otherwise: rules *without memory* or peripheral rules).

- *Legal*[7] rules: rules which do not change the null configuration[8] ("nothing comes of nothing").

Exercise 2.3.2. (**)
Cellular automata with $k = 2$, $r = 1$:

- How many rules are symmetric?
- How many rules are legal symmetric?
- How many rules are totalistic?
- How many rules have memory?

Exercise 2.3.3. (***)
Prove the following theorem: All legal symmetric rules of cellular automata with $k = 2$ and $r = 1$ form an additive group with elements 0, $f_0 = a_{i-1} + a_i + a_{i+1}$, $f_1 = a_{i-1} \cdot a_i + a_i \cdot a_{i+1} + a_{i-1} \cdot a_{i+1}$, $f_2 = a_{i-1} \cdot a_i \cdot a_{i+1}$, $f_3 = (a_i - a_{i-1})(a_i - a_{i+1})$, $f_4 = f_1 \cdot f_3$.

---

[7] Please note that some authors require that legal rules should also be symmetric.
[8] The null configuration is the (global) configuration where all cells are empty. In CA with legal rules it is also called the quiescent configuration.

Cellular automata as a discretization of partial differential equations? Lattice-gas cellular automata - a special type of cellular automata - are relatively new numerical schemes to solve physical problems ruled by partial differential equations. One could ask whether cellular automata can be interpreted as discrete models of partial differential equations. Consider the *diffusion equation*

$$\frac{\partial C}{\partial t} = \kappa \frac{\partial^2 C}{\partial x^2} \tag{2.3.3}$$

as an example of a partial differential equations of first order in time. The discretization forward in time and symmetric in space reads

$$C_i^{(t)} = C_i^{(t-1)} + \frac{\Delta t \cdot \kappa}{(\Delta x)^2} \left[C_{i+1}^{(t-1)} - 2C_i^{(t-1)} + C_{i-1}^{(t-1)} \right] \tag{2.3.4}$$

$$= \sum_{j=-1}^{j=1} \alpha_j C_{i+j}^{(t-1)}$$

$$= f \left[\sum_{j=-1}^{j=1} \alpha_j C_{i+j}^{(t-1)} \right]. \tag{2.3.5}$$

Here f is the identity. Eq. (2.3.5) is of the same form as the map (2.3.2) which defines the automata rule. However, there are fundamental differences:

- The coefficients α_j in (2.3.5) in general are real numbers and not integers.
- The number of states of C_i is infinite.
- In general C_i in Eq. (2.3.4) is not bounded whereas the result of $f()$ in Eq. (2.3.2) is limited to the range 0 to $k-1$ (modulo constraint).
- Whereas the development in time of the finite number of states is always stable the iteration of (2.3.4) can lead to instability, i.e. the absolute value of the concentration C_i goes to infinity (try to iterate (2.3.4) with a time step Δt below or slightly above the stability limit $\Delta t_c = \frac{(\Delta x)^2}{2\kappa}$).
- The diffusion equation (and many other partial differential equations in mathematical physics) are based on conservation laws whereas for most of the automata rules no conservation laws are known.

Although there are some formal similarities between discretization of partial differential equations and cellular automata rules the differences dominate. Only special types of cellular automata provide discrete models for partial differential equations of mathematical physics. The connection between the differential equations and lattice gas automata is not formal but deeply rooted in the ground of conservation laws.

Irreversibility and Garden of Eden configurations. *An important feature of (most) CA is their local irreversibility, i.e. under certain local rules different initial (global) configurations may be transformed into the same final configuration. As a consequence of irreversibility not all possible (global) configurations can be reached by time evolution of the CA. The unreachable configurations can only be initialized and therefore are called Garden of Eden configurations.*

Under most local rules cellular automata behave *locally irreversible*, i.e. different initial configurations are mapped onto the same final configuration. For deterministic rules each configuration has a definite post-configuration (descendant) which can result, however, from several initial configurations (ancestors). Hence the trajectories traced out by the time evolution of several configurations may coalesce, but may never split. A trivial example is provided by a CA with the totalistic null rule: the first iteration transforms arbitrary initial configurations into the null configuration. In a reversible system all configurations have definite post- and pre-configurations. Thus the number of accessible configurations is constant in time (*Liouville's theorem*) and is equal to the number of all possible configurations.

As a consequence of irreversibility there exist configurations that can be initialized but are unreachable during the development in time of the CA. Such configurations are called *Garden of Eden configurations* (Moore, 1962; Aggarwal, 1973). These configurations are not at all seldom. Under the null rule, for example, all configurations except for the null configuration lay in Paradise. Table 2.3.4 gives the fraction of reachable configurations for several rules of elementary cellular automata. Further investigations of Garden of Eden configurations can be found in Voorhees (1990, 1994, 1996), Voorhees and Bradshaw (1994) and Schadschneider and Schreckenberg (1998).

One of the basic decision problems of CA is to decide for a given local rule, whether its global rule has a Garden of Eden (Kutrib et al., 1997). It has been shown to be undecidable for two- and higher-dimensional CA (Kari, 1990; Durand, 1994) while it is decidable for one-dimensional CA (Amoroso and Patt, 1972).

Exercise 2.3.4. (**)
How many configurations of the cellular automata with $N = 10$, $k = 2$, $r = 1$, periodic boundary conditions, and rule 56 belong to the Garden of Eden?

Exercise 2.3.5. (**)
Which rules for cellular automata with $k = 2$, $r = 1$, periodic boundary conditions, and $N = 4$ or $N = 5$ are reversible?

The irreversible behavior of cellular automata is reflected also in the evolution in time of the information-theoretical (Shannon) entropy S which is defined

Table 2.3.4. *Reachable configurations of elementary cellular automata (k = 2, r = 1) with periodic boundary condions; compare Wolfram (1983).* $F_r \leq 1$ *is the fraction of reachable configurations (the number of all possible configurations is 2^N where N is the number of cells).*

Rule	F_r	Remarks
0	$1/2^N$	null rule is trivially irreversible
4	$1/2^{N-1}$	no two adjacent sites have the same value
90	$1/2$	if N is odd; even number of cells have value one
90	$1/4$	if N is even
126		depends on N; $\lim_{N \to \infty} F_r \to 0$
204	1	identity transformation is trivially reversible

as usual (but an arbitrary multiplicative constant or a different base for the logarithm[9] can be chosen) by

$$S := - \sum_i p_i \log_2 p_i \tag{2.3.6}$$

(see, for example, Wolfram, 1983)[10], whereby p_i is the probability of the (global) configuration i. The increase in entropy $S(t)$ with time (compare Exercise 2.3.8) is a reflection of local irreversibility of CA.

Exercise 2.3.6. (*)
Prove:
$$\lim_{x \to 0} x \log_2 x = 0$$

Exercise 2.3.7. (**)
Which distribution p_i belongs to an extremum of S?

Exercise 2.3.8. (**)
Calculate $S(t)$ for $t = 1$ to 100 for the CA with $k = r = 2$, periodic boundary conditions, $N = 10$ cells, and the totalistic rule 2. The initial ensemble encompasses all possible configurations with equal probabilities.

2.3.1 Qualitative characterization of one-dimensional cellular automata

The following rule numbers refer to legal totalistic rules with two states per cell $k = 2$ and range $r = 2$:

---

[9] The natural logarithm is more appropriate for calculations involving differentiation.

[10] Please note that Wolfram defines the entropy with a different sign: $S_W := + \sum_i p_i \log_2 p_i$ which actually gives the 'information content'.

$$a_i^{(t)} = f \left[\underbrace{\sum_{j=-2}^{j=2} a_{i+j}^{(t-1)}}_{=:\, s} \right] \tag{2.3.7}$$

The argument s can take on values between and 0 and 5 only[11]. Accordingly a rule is defined by six numbers $b_i \in \{0,1\}$. The sequence $b_5 b_4 b_3 b_2 b_1 b_0$ can be interpreted as the binary representation of an integer between 0 and 63 which refer to the various totalistic rules. Example: rule $20 \rightarrow b_5 b_4 b_3 b_2 b_1 b_0 = 010100$

$a_i^{(t)} = 1$ if $\sum_{j=-2}^{j=2} a_{i+j}^{(t-1)} = 2$ or 4 and

$a_i^{(t)} = 0$ otherwise.

Wolfram (1984a,b) has investigated a large number of one-dimensional automata with legal totalistic rules, two states per cell $k = 2$, range $r = 2$ and random initial conditions. He proposed the following classification[12] with four different types of behavior:

1. The final configuration is homogeneous.
   Rules: 0,4,16,32,36,48,54,60,62.
   Analogue in continuous dynamical systems: limit point.

2. The development in the course of time leads to simple time-independent or time-periodic patterns.
   Rules: 8,24,40,56,58.
   Analogue in continuous dynamical systems: limit cycles.

3. Generation of chaotic patterns.
   Rules: 2,6,10,12,14,18,22,26,28,30,34,38,42,44,46,50.
   Analogue in continuous dynamical systems: strange attractors.

4. The development in the course of time leads to complex local patterns which in part may be long-lived.
   Rules: 20,52.
   There is no analogue in continuous dynamical systems.

The following figures show the development in time of one-dimensional cellular automata with $k = 2$ possible states per cell, range $r = 2$, $N = 100$ or $N = 400$ number of cells, periodic boundary conditions, and legal totalistic rules. The initial configuration (upper line) is set randomly with equal probability to 0 (white) and 1 (black). All figures show the configurations at N consecutive time levels (from top to bottom).

---

[11] For arbitrary k and r: $0 \leq s \leq (k-1)(2r+1)$.

[12] Different classification schemes have been proposed by several authors (Stauffer, 1989; Gerling, 1990a; Binder, 1991; Twining 1992; Cattaneo et al., 1995; Makowiec, 1997).

Fig. 2.3.2. *Cellular automaton with $k = 2$ possible states per cell, range $r = 2$, $N = 400$ number of cells, periodic boundary conditions, and random initial configuration (upper line). The figure shows the configurations at 400 consecutive time levels (from top to bottom). The CA with totalistic rule 2 applies under Wolfram's third class. The connection between CA and Sierpinski carpets are discussed, for example, in Wolfram (1983) or Peitgen et al. (1992).*

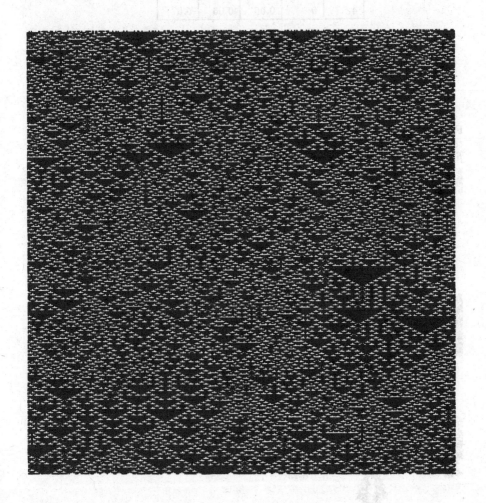

Table 2.3.5. *Legal totalistic cellular automata: Classification (approximately!) according to Wolfram (1984b).*

Type	$k = 2$ $r = 1$	$k = 2$ $r = 2$	$k = 2$ $r = 3$	$k = 3$ $r = 1$
1	0.5	0.25	0.09	0.12
2	0.25	0.16	0.11	0.19
3	0.25	0.53	0.73	0.60
4	0	0.06	0.06	0.07

Fig. 2.3.3. *1D CA with $k = 2$, $r = 1$, $N = 100$, periodic boundary conditions, totalistic rule 20 (Wolfram's class 4): the nine plots show the configurations at the first hundred time levels starting from different random initial configurations.*

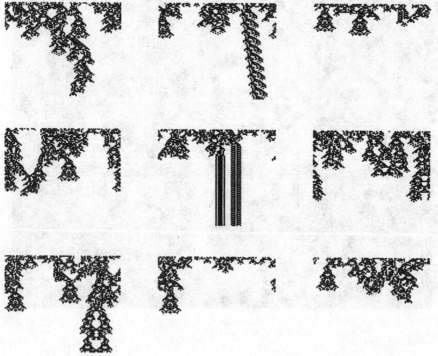

Fig. 2.3.4. *Same as Fig. (2.3.3) except totalistic rule 52 (Wolfram's class 4).*

Fig. 2.3.5. *1D CA with* $k = 2$, $r = 1$, $N = 100$, *periodic boundary conditions, totalistic rules* 2, 6, 10, 12, 14, 18, 22, 26, 28 *(from left to right and from top to bottom; Wolfram's class 3): the nine plots show the configurations at the first hundred time levels starting from the same random initial configuration.*

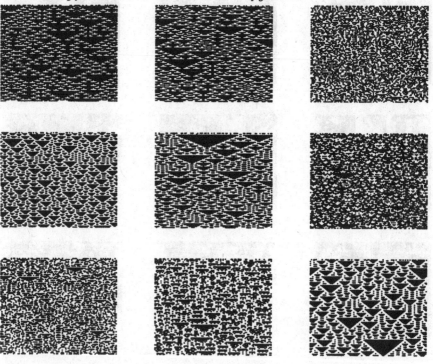

2.4 Two-dimensional cellular automata

In two dimensions there is much more freedom for arranging the cells and defining the neighborhoods for the updating rules. Here only the simplest configurations will be considered. Various other arrangements will be presented in the chapter on lattice-gas cellular automata.

2.4.1 Neighborhoods in 2D

Von Neumann neighborhoods of range r are defined by

$$N_{i,j}^{(vN)} := \left\{ (k,l) \in L \,\big|\, |k-i| + |l-j| \leq r \right\} \qquad (2.4.8)$$

and *Moore neighborhoods* of range r by

$$N_{i,j}^{(M)} := \left\{ (k,l) \in L \,\big|\, |k-i| \leq r \quad \text{and} \quad |l-j| \leq r \right\}. \qquad (2.4.9)$$

Fig. 2.4.6. *Neighborhoods in 2D of range 1 and 2: von Neumann (upper), Moore (lower).*

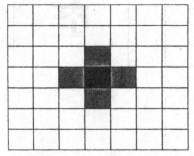
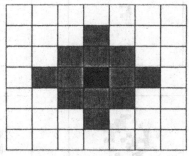

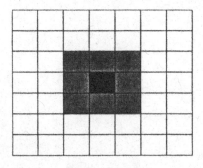
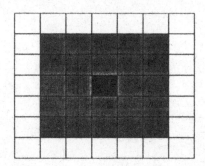

2.4.2 Fredkin's game

Fredkin proposed a cellular automata game with simple rules which leeds to self-replication in a trivial sense, i.e. without configurations that contain *universal Turing machines*. The game is defined as follows (Gardner, 1971b). Each cell has two possible states: alive (occupied) or dead (empty). All cells are updated simultaneously. Count the number of live cells of the four neighbors (von Neumann neighborhood of range 1; compare Fig. 2.4.6). Each cell with an even number $(0, 2, 4)$ of live neighbors will be dead at the next time level and alive otherwise. It can be shown that any initial pattern of live cells will reproduce itself four times after 2^n iterations (n depends on the initial pattern). The four replicas will be displaced 2^n cells from the vanished original. Fig. 2.4.7 shows an example where $n = 2$.

Fig. 2.4.7. *Fredkin's game: after 2^n iterations the original pattern of live cells has disappeared and four replicas have shown up at a distance of 2^n cells from the vanished original. n depends on the initial pattern and is equal to 2 in the example shown here.*

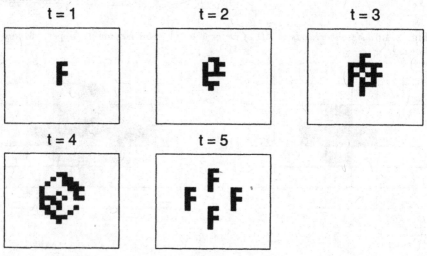

2.4.3 'Life'

At the beginning of the 70's Conway introduced the *'Life'*: a two-dimensional synchronous cellular automaton which simulates the evolution of a society of living organisms.
'Life' is defined by two rules involving eight neighbors (Moore neighborhood of range 1; compare Fig. 2.4.6):

– Each live site will remain alive the next time-step if it has two or three live neighbors, otherwise it will die.
– At a dead site new live will be born only if there are exactly three live neighbors.

'Life' contains many patterns which remain stable from iteration to iteration when not disturbed by other objects (see Fig. 2.4.8 for some examples).

Fig. 2.4.8. *The patterns shown here remain stable from generation to generation.*

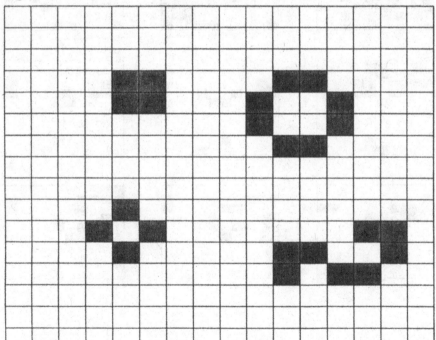

The development in time of initial random configurations with equal probabilities of 1/2 for dead or alive is shown in Figures (2.4.9) and (2.4.10). In the limit of large domain size and time approximately 3 % of all cells are alive (compare Fig. 2.4.11).

Fig. 2.4.9. *'Life' on a* 10 *times* 10 *array with periodic boundary conditions. The figure shows the random initialization with equal probability for dead or alive (upper left) and the configurations at the eight successive time levels (from left to right and downward).*

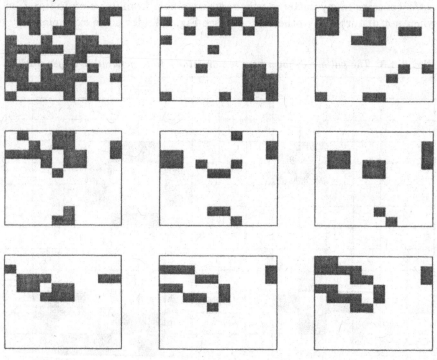

Fig. 2.4.10. *'Life' on a 50 times 50 array with periodic boundary conditions. Upper left: Random initialization with equal probability for dead or alive. The other plots show the configurations at time levels 141 to 148.*

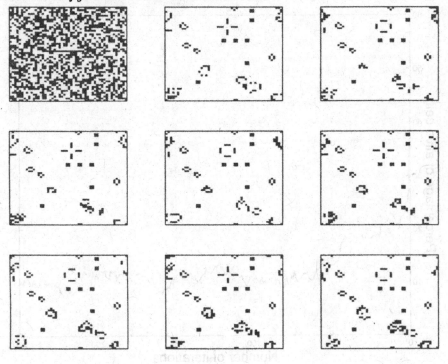

Fig. 2.4.11. *'Life' on a 50 times 50 array with periodic boundary conditions: percentage of alive cells as a function of time (iterations). The limit for large domain size and time is not yet reached.*

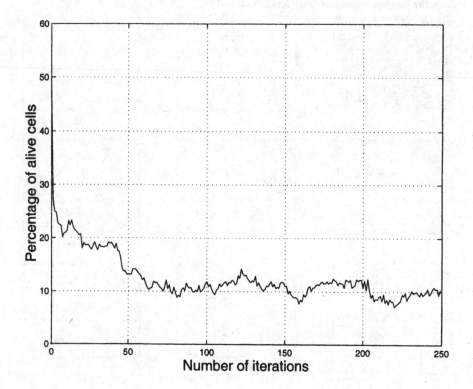

Exercise 2.4.1. (**)
Find all stable configurations with an extension of 3 times 3 cells at most.

Exercise 2.4.2. (*)
The rule of Conway's *'Life'* is one of many possible rules for a two-state cellular automaton in two-dimensions when the rule takes into account the states of the nearest neighbors and the cell itself. One can wonder whether the self-organized critical regime can occur for *'Life'* only or can occur for many other of the possible rules. Suppose a supercomputer needs only one microsecond to investigate one single rule. How long does it take to analyse all the possible rules?

Further reading:
Sigmund (1993), Alstrøm and Leão (1994), Vandewalle and Ausloos (1995), Bak (1996), Heudin (1996), Pulsifer and Reiter (1996), Nordfalk and Alstrøm (1996), de la Torre and Martin (1997), Malarz et al. (1998).

2.4.4 CA: what else? Further reading

Books, proceedings, reviews:

- Codd (1968)
- Burks (1970)
- Wolfram (1986)
- Jackson (1990, chapter 10)
- Gutowitz (1991),
- Wolfram (1994)
- Vollmar et al. (1996)
- Voorhees (1996)
- Kutrib et al. (1997)
- 2nd Conference on Cellular Automata for Research and Industry, Theoretical Computer Science, 217(1), 1999.

Articles:

- Invertible CA: Toffoli and Margolus (1990).
- Elementary reversible CA: Takesue (1987, 1989, 1990).
- Additive invariants: Hattori and Takesue (1991).
- Staggered invariants: Takesue (1995).

- Normal forms of CA: see references in Kutrib et al. (1997).

- Fractals: Peitgen et al. (1998).

- Simulation of traffic flow with cellular automata: Nagel and Schreckenberg (1992), Schreckenberg et al. (1995), Benjamin et al. (1996), Brankov et al. (1996), Chopard et al. (1996), Fukui and Ishibashi (1996a,b), Ishibashi and Fukui (1996a,b), Kerner et al. (1996), Rickert et al. (1996), Barlovic et al. (1998), Emmerich et al. (1998), Huang (1998), Nagel et al. (1998), Schadschneider and Schreckenberg (1998), Wang et al. (1998).

- Biology: Stuart A. Kauffman (1984, 1986, 1991) applied CA with complex interconnection to biological problems; Ermentrout and Edlestein-Keshet (1993) give a review of biologically motivated CA.
  Bandini et al. (1998), Bastolla and Parisi (1998a,b), Mielke and Pandey (1998), Siregar et al. (1996, 1998), Zorzenon Dos Santos (1998), O'Toole et al. (1999).

- Statistical mechanics: Rujàn (1987)

- Pattern formation: Lindgren et al. (1998).

- Quantum cellular automata: Watrous (1995), Richter and Werner (1996), Tougaw and Lent (1996)

- More papers can be found in the following journals:

  - Complex Systems

  - Physica D (for example: volume 45, p. 3-479, 1990 and volume 103, 1997).

2.4.5 From CA to LGCA

Despite of their simple update rules cellular automata can display complex behavior which is a prerequisite to use them as a simulation tool for physical (biological, chemical, ...) phenomena like, for example, fluid flow. CA are very easy to implement and are especially well suited for massively parallel computers because of the local character of the update rules. By construction they are unconditionally numerically stable.

Before CA are to be used for the simulation of physical processes the following items have to be addressed:

1. Many physical laws are based on the conservation of certain quantities. The Navier-Stokes equation, for example, expresses the conservation of mass and momentum. The cellular automata used for simulation should hold corresponding conserved quantities. As will be shown later on one of the main problems in the construction of lattice-gas cellular automata is to avoid the occurence of additional (non-physical or spurious) invariants.

2. Nonstatic physical phenomena involve the transport of certain quantities. The propagation of information in CA is not possible for most rules. Among the elements of the group $G = (0, f_0, f_1, f_2, f_3, f_4; +)$ introduced in Exercise 2.3.3, for example, only one special element (f_0) can generate propagation.

3. The desired physical behavior of a lattice-gas cellular automata will show up in the macroscopic limit which can be derived from a theory of statistical mechanics on a lattice. The application of certain concepts of statistical mechanics requires that the microdynamics, i.e. the update rules, are *reversible*.

Only a small subset of CA holds the appropriate number of conserved quantities, is able to propagate these quantities or has reversible rules. Based on the discussion of CA given above it is not clear whether or not CA exist with all three properties and if so, how to construct such CA for a given phenomenon.

In order to simplify the problem of constructing cellular automata for given physical processes lattice-gas cellular automata (LGCA) differ somewhat from the CA discussed above in that the update is split into two parts which are called *collision* and *propagation* (or *streaming*). The collision rule of LGCA can be compared with the update rule for CA in that it assigns new values to each cell based on the values of the cells in a local neighborhood. After the collision step the state of each cell is propagated to a neighboring cell. This split of the update guarantees propagation of quantities while keeping the proper update rules (collisions) simple. Because of this difference of LGCA from the CA discussed here so far some authors do not consider LGCA as proper CA and prefer to speak of *lattice gases* (see, for example, Hénon, 1989b). This latter notation might, however, lead to confusion. "There are at least two, almost independent, lattice gas communities: one community ... usually moves bits or numbers around a lattice while conserving momentum; the other community, mostly solid-state theorists, focuses almost exclusively on the Ising model." (Quoted from J. Stat. Phys., Vol. 68, Nos. 3/4, p. 611, 1992.)

3. Lattice-gas cellular automata

3.1 The HPP lattice-gas cellular automata

The first lattice-gas cellular automata (LGCA) was proposed in 1973 by Hardy, de Pazzis and Pomeau. It is named HPP after the initials of the three authors. The HPP model is the simplest[1] LGCA which will be discussed in some length. Today HPP is of interest mainly for historical reasons[2] because it does not lead to the Navier-Stokes equation in the macroscopic limit. In the current chapter specific coding techniques for lattice-gas cellular automata[3] like multi-spin coding will be introduced.

3.1.1 Model description

HPP is a two-dimensional lattice-gas cellular automata model over a square lattice. The vectors c_i ($i = 1, 2, 3, 4$) connecting nearest neighbors (compare Fig. 3.1.1) are called *lattice vectors* or *lattice velocities*. More precisely, the lattice velocities are given by the lattice vectors divided by the time step Δt which is always set equal to 1. So lattice vectors and lattice velocities have different dimensions but the same numerical values. The meaning of the c_i can be easily recognized from the respective context. At each *site* (*node*) there are four cells (Fig. 3.1.2) each associated to a link with the nearest neighbor. These cells may be empty or occupied by at most one particle. This *exclusion principle* (*Pauli principle*) is characteristic for all lattice-gas cellular automata. It will lead to equilibrium distributions of *Fermi-Dirac*[4]

---

[1] The model of Boghosian and Levermore (1987) for Burger's equation in one spatial dimension will not be discussed here.

[2] There are still some applications: Chopard and Droz (1991) use HPP as a random generator.

[3] There are only few papers which give hints to specific coding techniques for lattice-gas cellular automata; see, for example, Kohring (1991), Wolf-Gladrow and Vogeler (1992), and Slone and Rodrigue (1997).

[4] Fermi-Dirac distributions are well known from quantum mechanics. Particles with half-odd-integer spins like the electron, proton, or neutron are called fermions; they obey Fermi-Dirac distributions.

type for the mean occupation of the cells. All particles have the same mass m (which will be set to 1 for simplicity) and are indistinguishable.

The evolution in time is deterministic and proceeds as an alternation of local *collisions* C (only particles at the same node are involved) and *streaming* S (also called *propagation*) along the appropriate links to the nearest neighbors. The *evolution operator* $\mathcal{E}$ is defined as the composition of collision and streaming:

$$\mathcal{E} = S \circ C. \tag{3.1.1}$$

To each particle a momentum of magnitude mc_i is assigned. The collision should conserve mass and momentum while changing the occupation of the cells. For HPP there is only one collision configuration. When two particles enter a node from opposite directions and the other two cells are empty a *head-on collision* takes place which rotates both particles by 90° in the same sense (compare Fig. 3.1.3). All other configuration stay unchanged during the collision step. In passing we note that twofold application of the collision operator leads back to the initial configuration:

$$C^2 = \mathcal{I}, \tag{3.1.2}$$

where $\mathcal{I}$ is the *identity operator*.

The HPP model respects a particle-hole symmetry, i.e. the operator $\mathcal{F}$ - which interchanges particles and holes - commutes with the evolution operator $\mathcal{E}$. As a consequence of this symmetry the model has similar properties at low and corresponding high mass densities (Hardy et al., 1973).

At each time step particles are interchanged between the sub-lattice consisting of points with even indices (the 'white' sub-lattice) and the sub-lattice consisting of points with odd indices (the 'black' sub-lattice; imagine a chessboard). Therefore there exist two decoupled particle populations on the lattice. This decoupling is characteristic for the square lattice (compare the decoupling of solutions at even and odd time steps - sometimes called the *chessboard instability* - in finite differences; see, for example, Orszag, 1971 or Rood, 1987).

As already noted above the HPP model does not obey the desired hydrodynamic equations (Navier-Stokes) in the macroscopic limit. We will prove later on that this deficit is due to the insufficient degree of rotational symmetry of the lattice. Certain tensors composed of products of the lattice velocities - so-called *lattice tensors* - are not isotropic over the square lattice. See Section 3.3 for an extensive discussion of these tensors. This anisotropy would manifest itself, for example, in the flow past a non-rotational symmetric obstacle in that the drag depends on the relative orientation of the obstacle with respect to the lattice.

In addition to mass and momentum there exist additional conserved quantities for the HPP model. For example, the difference in the number of particles parallel and anti-parallel to a lattice axis does not change by collisions

Fig. 3.1.1. *The square lattice of the HPP model. The four arrows labelled by a, b, c, and d indicate the lattice velocities c_i. Particles are interchanged between the black (small points) and the white (small circles) sub-lattices (chess board).*

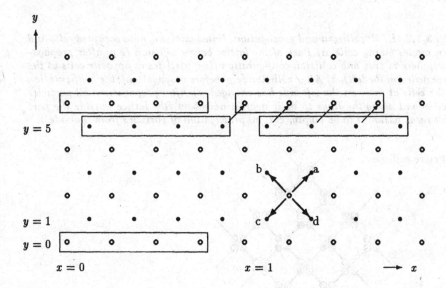

or propagation. These '*spurious invariants*' are undesirable because they restrict to a certain degree the dynamics of the model and have no counterpart in the real world.

Fig. 3.1.2. *HPP: collision and propagation. Filled circles denote occupied cells and open circles empty cells. a) Part of the lattice before collision (e.g. after propagation); there is only one collision configuration (two particles in opposite cells at the same node; on the left). b) After collision (e.g. before propagation): the configuration of the cells at node on the left side has changed. c) After propagation: all particles have moved along the links to their nearest neighbors (the lattice outside the part shown was assumed to be empty, e.g. no propagation of particles from 'outside').*

a) before collision

b) after collision

c) after propagation

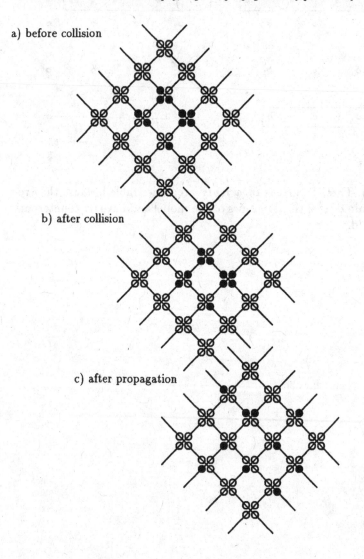

Fig. 3.1.3. *HPP: collisions. a) There is only one collision configuration (head-on collision) for HPP: two cell on opposite links are occupied and the two other cells are empty. After collision the formerly empty cells are occupied and vice versa. b) Same as a) but showing the associated momentum vectors. Both momentum vectors are rotated by 90°. Mass and momentum are conserved.*

a)

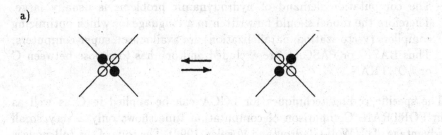

b)

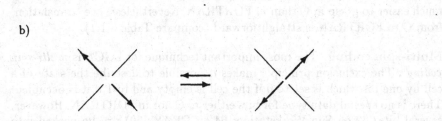

3.1.2 Implementation of the HPP model: How to code lattice-gas cellular automata?

The coding techniques discussed here will be also applicable to other LGCA like FHP or PI. The FCHC model requires a different approach (discussed in Section 3.5).

FORTRAN or C? Which programming language is best suited for the coding of LGCA? Of course there will be no unique answer to this question and often discussions with various people resemble religious controversies[5]. I propose the following criteria:

- The computer codes should be portable, i.e. programming in machine specific language (Assembler) is excluded.

- The computation demand of hydrodynamic problems is usually large. Therefore the model should be written in a language for which optimizing compilers (vectorization, parallelization) are available on super-computers. Thus BASIC or PASCAL are excluded and one has to choose between C or FORTRAN.

The specific coding techniques for LGCA can be applied in C as well as in FORTRAN. Comparison of computation time shows only a very small advantage of C (Wolf-Gladrow and Vogeler, 1992). The code of the collisions is much easier to grasp in C than in FORTRAN. Nevertheless, the 'translation' from C to FORTRAN is straightforward (compare Table 3.1.1).

Multi-spin coding. The most important technique for LGCA is *multi-spin coding*[6]. The exclusion principle makes it possible to describe the state of a cell by one bit which is set to 0 if the cell is empty and to 1 if it is occupied. There is no special data type for bits either in C nor in FORTRAN. However, several bits (32 on Sun-Workstation; 64 on CRAY-J90) can be packed into one unsigned (C) or integer (FORTRAN) variable. In standard C there exist bit-operators which act bitwise[7] on whole unsigned variables. In FORTRAN you may find bit-functions with the same effects. These bit-functions do not belong to the FORTRAN standard but are available on almost all machines.

The core of the HPP program, namely the coding of collision and streaming, encompasses only a few lines. Here is the code in C:

---

[5] I was taught that 'FORTRAN is a dead language' already in the 70ies.

[6] Although the term 'multi-spin coding' which is related to the spins of Ising models has been coined by Creutz et al. in 1979, this technique has been described already in the article by Hardy et al. (1976) and it was first mentioned by Friedberg and Cameron (1970).

[7] Example: decimal 65 | 39 reads in binary notation on an 8-bit machine (01000001) | (00010111) = (01010111) and therefore 65 | 39 = 103.

Table 3.1.1. *Some elements and constructions in C and FORTRAN.*

C	FORTRAN	Remarks
a & b	iand(a,b)	bit operators are
a | b	ior(a,b)	standard in C;
a ^ b	ieor(a,b)	bit functions are
~a	not(a)	(almost always available)
		extensions in FORTRAN
a<<3	ishift(a,3)	left shift of bits by
		3 positions
a>>4	ishift(a,-4)	right shift
a = a | (1<<3)	a = ibset(a,4)	set 4th bit
#define A 5	parameter (A=5)	in C: global define
unsigned a[3]	integer a(3)	1D array with 3 elements
a[0],a[1],a[2]	a(1),a(2),a(3)	elements
for(i=0;i<3;i++){	do i=1,3	loops
}	enddo	
a % b	mod(a,b)	a modulo b

Table 3.1.2. *Bit-operators in C: & and; | inclusive or; ∧ exclusive or; ~ not.*

a	b	a & b	a | b	a ∧ b	~ a
0	0	0	0	0	1
1	0	0	1	1	0
0	1	0	1	1	1
1	1	1	1	0	0

```
/* ---  grid   ---

    b       a
     \     /
      \   /
       \ /
        +
       / \
      /   \
     /     \
    c       d           */

    last = LENGTH - 1;    /* --- shift last bit --- */
    XMAX1 = XMAX - 1;
    YMAX1 = YMAX - 1;

                         /* -----  collision  ----- */
for(x=0; x<XMAX; x++)
for(y=0; y<YMAX; y++) {
change = ( (a1[x][y] & c1[x][y] & ~(b1[x][y] | d1[x][y])) |
           (b1[x][y] & d1[x][y] & ~(a1[x][y] | c1[x][y])) )
         & nsb[x][y];
a2[x][y] = a1[x][y] ^ change;
b2[x][y] = b1[x][y] ^ change;
c2[x][y] = c1[x][y] ^ change;
d2[x][y] = d1[x][y] ^ change; }

/* -----   propagation in a-direction   ------ */

for(x=1; x < XMAX1; x++) {
for(y=1; y < YMAX1; y += 2) {

/* note: brackets are necessary,
   because in C the + has a higher priority than the >>  */

/* black to white: */
a1[x][y]   = (a2[x][y-1] >> 1) + (a2[x-1][y-1] << last);
a1[x][y+1] = a2[x][y]; }}        /* white to black: */
```

A few comments on the code are now in order:

– Each unsigned variable can store $LENGTH$ (= 64 on CRAY computers) bits. The number of grid points is $XMAX*LENGTH$ in x-direction and $YMAX$ in y-direction (compare Fig. 3.1.1).

- The states of all cells are stored in the two-dimensional (2D) arrays $a1$, $b1$, $c1$, $d1$, where a, b, c, d assign the different lattice directions (lattice velocities). $a1[0][0]$ contains the 'a-bits' of the nodes from 1 to $LENGTH$ of the first line.

- The location of obstacles is stored in the 2D array nbs (non-solid bit): bits in nbs are set to 1 outside of the obstacles and 0 otherwise.

- In the first loop a variable $change$ is calculated. It contains the information whether or not a collision will happen: the bits in $change$ are set to 1 if the configurations $(a, b, c, d) = (1010)$ or (0101) are present and the node is located outside of obstacles.

- Subsequently, the bit arrays $a1, b1, c1, d1$ are concatenated by 'exclusive or' with the variable $change$. This changes the state bits (interchange of 0 and 1; compare Table 3.1.2). The results of this operation are stored in the auxiliary arrays $a2, b2, c2, d2$. The introduction of these auxiliary arrays ensures a very fast updating on vector computers. These arrays are also useful in the propagation step.

- In Fig. 3.1.1 the propagation is shown only for the a-directions (upwards and to the right). In the program listing the propagation in a-direction is shown only for the inner nodes. The propagation for nodes on the boundaries has to be treated separately according to the appropriate boundary conditions.

  - The propagation from the white to the black sub-lattice consists of a storage in different arrays.

  - The propagation from the black to the white sub-lattice is more involved. To simplify the following discussion let us assume that number of bits per integer is only 4 (the parameter $LENGTH$ in the code). The propagation of $a2[0][5]$ and $a2[1][5]$ to $a1[1][6]$ (compare Fig. 3.1.1) will be considered. First all bits of $a2[1][5]$ will be shifted to the right whereby the rightmost bit drops out. The resulting void at the left boundary of $a2[1][5]$ will be filled up automatically by a 0. Yet, this position must be occupied by the rightmost bit of the neighbor element $a2[0][5]$. To isolate this bit all bits in $a2[0][5]$ are shifted to the left by $LENGTH$-1 (= 'last' in the code) digits whereby zeros fill up from the right. Now $a2[1][5]$ after a right shift and $a2[0][5]$ after $LENGTH$-1 left shifts could be concatenated by exclusive or. But the addition of these two shifted integers yields the same result and is faster on some computers due to chaining[8] (compare Kohring, 1991 and Wolf-Gladrow and Vogeler, 1992).
  Example: $LENGTH = 4$; $last = 3$;

---

[8] Chaining is the process of passing the output of one vector operation directly as input into another vector operation. As soon as the first element of the first operation's result is output, the second operation can begin. This allows partial overlapping of vector instruction execution.

a2[0][5] = (1001); a2[1][5] = (1011);
a2[1][5] >> 1 = (0101); a2[0][5] << last = (1000)
a1[1][6] = (1101)

3.1.3 Initialization

Before the evolution of the LGCA can start the various arrays have to be initialized. At time $t = 0$ the bits are set by random processes with probabilities such that the mean values over a large number of nodes (typically 32 times 32 or 64 times 64) approximate the given initial values for mass and momentum density. Thus the question arises, how to choose appropriate probabilities for given mass and momentum density?

The state of the LGCA is fully described by the Boolean fields $n_i(t, r_j)$ where the index i which runs from 1 to 4 (or alternatively from a to d) indicates the directions, n_i is the *occupation number* which may be 0 or 1, t is the (discrete) time and r_j are the coordinates of the nodes. Mean occupation numbers N_i are calculated by averaging over neighboring nodes

$$N_i(t, x) = \langle n_i(t, r_j) \rangle . \tag{3.1.3}$$

The mean occupation numbers can take on values between 0 and 1. Mass $\rho(t, x)$ and momentum density $j(t, x)$ are defined by

$$\rho(t, x) = \sum_{i=1}^{4} N_i(t, x) \tag{3.1.4}$$

and

$$j(t, x) = \rho u = \sum_{i=1}^{4} c_i N_i(t, x) \tag{3.1.5}$$

(u is the flow velocity) with the lattice velocities c_i

$$c_1 = \frac{1}{\sqrt{2}}(1, 1) \tag{3.1.6}$$

$$c_2 = \frac{1}{\sqrt{2}}(-1, 1)$$

$$c_3 = \frac{1}{\sqrt{2}}(-1, -1) \tag{3.1.7}$$

$$c_4 = \frac{1}{\sqrt{2}}(1, -1)$$

which obey

$$\sum_{i=1}^{4} c_i = 0 \quad \text{(lattice symmetry!)} \tag{3.1.8}$$

and

$$\sum_{i=1}^{4} c_{i\alpha} c_{i\beta} = 2\delta_{\alpha\beta},\qquad(3.1.9)$$

where the Latin indices refer to the lattice vectors and run from 1 to 4 whereas the Greek indices assign the cartesian components of the vectors and therefore run from 1 to 2. This convention will be used also in all other chapters.

The theoretical background for the calculation of the equilibrium occupation numbers will be developed not until the next chapter (the section on the FHP model contains some results relevant for HPP). Instead a simple ansatz for N_i will be made which is linear in ρ and j

$$N_i = \xi\rho + \eta c_i j. \qquad(3.1.10)$$

The coefficients ξ and η can be calculated from the constraints (3.1.4) and (3.1.5):

$$
\begin{aligned}
\rho &= \sum_{i=1}^{4} N_i \\
&= \sum_i \xi\rho + \eta j \underbrace{\sum_i c_i}_{=0} \\
&= 4\xi\rho
\end{aligned}
\qquad(3.1.11)
$$

which yields $\xi = 1/4$;

$$
\begin{aligned}
j &= \sum_{i=1}^{4} c_i N_i \\
&= \xi\rho \underbrace{\sum_i c_i}_{=0} + \sum_i \eta c_i(c_i j) \\
&= 2\eta j
\end{aligned}
\qquad(3.1.12)
$$

thus $\eta = 1/2$ and therefore

$$N_i = \frac{\rho}{4} + \frac{1}{2}c_i j. \qquad(3.1.13)$$

I.e., for $j = 0$ the occupation numbers are independent of direction (they can still depend on location which corresponds to pure density perturbations) whereas non-vanishing momenta imply occupation numbers which vary with direction.

The Boolean arrays n_i will be initialized with probabilities[9] p_i such that the N_i give the desired distributions of ρ and j when summed up according to Eqs. (3.1.4) and (3.1.5). The relations between the Boolean arrays n_i and the mass and momentum density are illustrated in Fig. 3.1.4.

Fig. 3.1.4. *Relations between microscopic (Boolean arrays n_i) and macroscopic (mass and momentum density) level.*

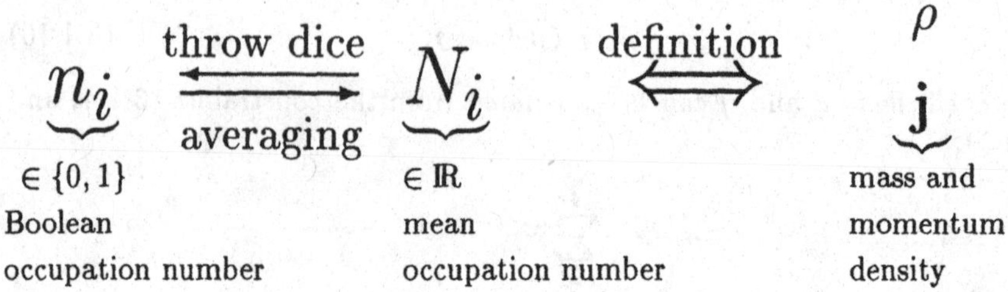

$$\underbrace{n_i}_{} \xrightleftharpoons[\text{averaging}]{\text{throw dice}} \underbrace{N_i}_{} \overset{\text{definition}}{\Longleftrightarrow} \begin{matrix} \rho \\ \underbrace{j}_{} \end{matrix}$$

$\in \{0, 1\}$	$\in \mathbb{R}$	mass and
Boolean	mean	momentum
occupation number	occupation number	density

Exercise 3.1.1. (*)
Construct occupation numbers N_i which vary with direction but yield $j = 0$.

Exercise 3.1.2. (**)
How long does it take for the distribution (3.1.13) to relax toward equilibrium distribution? What does the relaxation time constant depend on?

3.1.4 Coarse graining

The calculation of mean values for mass and momentum density is called coarse graining. Although it is possible to average over space, time or a combination of space and time, spatial coarse graining is much faster than the

---

[9] Unfortunately there is no standard for random generators. Portable random generators can be found for example in 'Numerical Recipes' (Press et al., 1992a,b). Different types of random generators (multiplicative congruential, shift-register and lagged Fibonacci) are discussed by Slone and Rodrigue (1997).

other alternatives. For the purpose of coarse graining the domain is divided into a number of subdomains which are large enough (usually 32 times 32 or 64 times 64 nodes) to obtain reliable (low noise) averages and small enough as to allow a large 'physical domain' under the constraint of a given limit of core memory.

The following more technical notes can be skipped in a first reading. The main computational load is the counting of the 1-bits in the unsigned (integer) arrays. On some computers a fast routine for counting the 1-bits in an integer is available. On the CRAY, for example, this is the FORTRAN function POPCNT (population count). Nothing similar is available in C. But one can use the FORTRAN function in C. Include the following lines into the code

```
#include <fortran.h>
fortran int POPCNT();      /* counts the number of 1-bits
                              in a 64 bit word */
```

and apply POPCNT to unsigned variables:

```
int n;
unsigned u;

u = 7;
n = POPCNT(u);
```

If no machine specific population count is available one can use the following routine which applies a *look-up table* (see Kohring, 1991, for the FORTRAN version):

```
/* -------- popcount for 32-bit unsigned ------ */

unsigned lu16[65536];    /* global */

void makelu16()
{
/* ----- make look-up table ----- */
    int ilu,ibi;
    for(ilu=0; ilu<65536; ilu++) {
        lu16[ilu] = 0;
        for(ibi=0; ibi<16; ibi++) {
```

```
                if( (ilu & bits[ibi]) > 0) lu16[ilu] += 1;
        }
    }
}   /* ---    end of makelu16   --- */

int popcount(unsigned u)
{
/* --- count bits --- */
    int pop;
    pop = lu16[(u&65535)] + lu16[( (u>>16)&65535 )];
    return(pop);
}   /* --- end of popcount --- */
```

3.2 The FHP lattice-gas cellular automata

In 1986 Frisch, Hasslacher and Pomeau showed that a lattice-gas cellular automata model over a lattice with a larger symmetry group than for the square lattice yields the incompressible Navier-Stokes equation in the macroscopic limit. This model with hexagonal symmetry is named FHP according to the initials of the three authors. The discovery of the symmetry constraint was the start for a rapid development of lattice-gas methods. The theoretical foundations where worked out by Wolfram (1986) and by Frisch et al. (1987). Within the following years many extensions and generalizations (FCHC for 3D simulations, colored models for miscible and immiscible fluids) were proposed. These models allow a wide range of applications.

After concentrating more on coding techniques in the HPP chapter the focus of the current section will be on the *theory* of lattice-gas cellular automata. Especially the equilibrium distribution and the macroscopic equations will be derived.

3.2.1 The lattice and the collision rules

The FHP lattice is composed of triangles (compare Fig. 3.2.1). It is invariant under rotations by $n \cdot 60°$ modulo $360°$ (*hexagonal symmetry*) about an axis through a node and perpendicular to the lattice plane. At each node and each link to the nearest neighbor there is a cell which may be empty or occupied by at most one particle (*exclusion principle*). All particles have the same mass m (set to 1 for simplicity) and are indistinguishable. The state of a node can be described by six bits. The exclusion principle leads to an equilibrium distribution of the mean occupation numbers of *Fermi-Dirac* type. Each cell is associated with a *lattice vector* c_i which connects a node with its nearest neighbor in direction i. The lattice vectors c_i are also called *lattice velocities* because the time step Δt is always set to 1 in lattice-gas cellular automata and therefore c_i and $c_i/\Delta t$ have the same numerical values. Because all particles have the same mass $m = 1$, c_i is also the particle's momentum. The respective meaning of c_i can be recognized from the context.

As for HPP there are 2-particle *head-on collisions* (compare Fig. 3.2.2). The initial state[10] $(i, i + 3)$ can be transformed into one of two different final states $(i + 1, i + 4)$ or $(i - 1, i + 2)$ (rotation by $60°$ to the left or right) while conserving mass and momentum density. If one chooses always one and the

---

[10] The description of the state of a node will be given in terms of indices of the occupied cells whereby the cell index $i > 0$ is understood as modulo 6. The index $i = 0$ will be assigned to rest particles (see below).

same final state the model becomes *chiral*[11]: it is not invariant with respect to spatial reflections (parity transformation). This is an undesired property because the hydrodynamic equations do not break parity symmetry. To restore reflection symmetry on the macroscopic level the choice between the different final states will be made by a random process with equal probabilities for rotation to the left and to the right. Thus in contrast to HPP the FHP model encompasses *nondeterministic rules*.

The generation of random numbers is a time consuming process. Therefore a *pseudo-random choice* is used where the rotational sense changes by chance for the whole domain from time step to time step (i.e. only one random number per time step has to be generated) or the sense of rotation changes from node to node but is constant in time (i.e. random numbers have to be generated only in the initial step for all nodes).

The 2-particle collisions conserve not only mass and momentum but also the difference of the number of particles that stream in opposite directions (the same invariant as for HPP). This additional invariant has no counterpart in 'real world hydrodynamics' and therefore is called a *spurious invariant*. It further restricts the dynamic of the lattice-gas cellular automata and can lead to deviations from hydrodynamic behavior on the macroscopic scale. The invariance of the particle differences can be destroyed by *symmetric 3-particle collisions* which conserve mass and momentum (compare Fig. 3.2.2). 2- and 3-particle collisions form a minimal set of collisions for FHP. This version of FHP is called *FHP-I*. Introduction of additional collisions like 4-particle collision, 2-particle collision with spectator and collisions including rest particles lead to various variants (*FHP-II, FHP-III*: see, for example, Frisch et al., 1987 and Hayot and Lakshmi, 1989). The corresponding macroscopic equations all have the same form (*universality theorem*) and differ only in their viscosity coefficients. As a rule of thumb the viscosity coefficient decreases with increasing number of collisions.

The 3-particle collisions destroy a spurious invariant. Unfortunately no method exists to detect all invariants of a given lattice-gas cellular automata model. Of course this is an unsatisfactory situation in the light that the invariants play an essential role in the equilibrium distributions of the mean occupation numbers. Indeed Zanetti (1989) found *spurious invariants* for all variants of the FHP model. Fortunately these *staggered invariants* are not set to values above a certain noise level by the usual initialization procedure and obviously are not generated by interactions with obstacles. So they do not influence the macroscopic dynamic too much. A discussion of the *Zanetti invariants* will be given in Section 3.8.

Frisch et al. (1987) give a discussion of a unified theory for the lattice-gas cellular automata HPP, FHP and FCHC. This excellent paper may be heavy fare for the beginner. The following discussion is restricted to the FHP model

---

[11] The word *chiral* derives from the Greek word for 'hands'.

Fig. 3.2.1. *The triangular lattice of the FHP model shows hexagonal symmetry. The lattice velocities c_i are represented by arrows.*

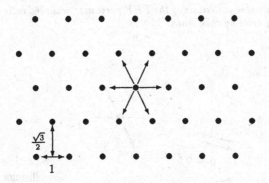

without rest particles (FHP-I) but from time to time more general results from Frisch et al. (1987) will be quoted.
The essential properties of the FHP model read:

1. The underlying regular lattice shows *hexagonal symmetry*.

2. *Nodes* (also called *sites*) are linked to six nearest neighbors located all at the same distance with respect to the central node.

3. The vectors c_i linking nearest neighbor nodes are called *lattice vectors* or *lattice velocities*

$$c_i = \left(\cos\frac{\pi}{3}i, \sin\frac{\pi}{3}i\right), \quad i = 1, ..., 6. \qquad (3.2.1)$$

   with $|c_i| = 1$ for all i.

4. A cell is associated with each link at all nodes.

5. Cells can be empty or occupied by at most one particle (*exclusion principle*).

6. All particles have the same mass (set to 1 for simplicity) and are indistinguishable.

7. The evolution in time proceeds by an alternation of *collision C* and *streaming S* (also called *propagation*):

$$\mathcal{E} = \mathcal{S} \circ \mathcal{C}, \qquad (3.2.2)$$

   where $\mathcal{E}$ is called the *evolution operator*.

8. The collisions are strictly *local*, i.e. only particles of a single node are involved.

Fig. 3.2.2. *All possible collisions of the FHP variants: occupied cells are represented by arrows, empty cells by thin lines.*

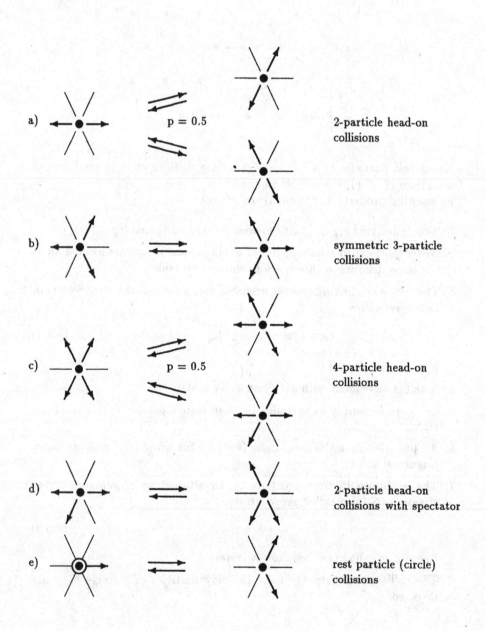

r is the position vector of a node and $r + c_i$ are the position vectors of its nearest neighbors. Each pair of lattice vectors (c_i, c_j) is associated with an element of the isometric group[12] $\mathcal{G}$ which transforms c_i into c_j by a rotation of $n \cdot 60°$. The first three moments of the lattice velocities c_i read:

$$\sum_i c_i = 0 \quad \text{(symmetry of the lattice!)} \tag{3.2.3}$$

$$\sum_i c_{i\alpha} c_{i\beta} = 3\delta_{\alpha\beta} \tag{3.2.4}$$

$$\sum_i c_{i\alpha} c_{i\beta} c_{i\gamma} = 0 \tag{3.2.5}$$

where the sum over i always runs from 1 to 6. The *Latin indices* refer to the lattice vectors and run from 1 to 6 whereas the *Greek indices* assign the cartesian components of the vectors and therefore run from 1 to 2. To each node a *bit-state* $n(r) = \{n_i(r), i = 1, ..., 6\}$ will be assigned where the $n_i \in \{0, 1\}$ are Boolean variables. The *streaming (propagation)* is defined by

$$n_i(r) = \mathcal{S} n_i(r - c_i). \tag{3.2.6}$$

Collisions take place synchronously at every node and transform the *initial state* of a node $s = \{s_i, i = 1, ..., 6\}$ into the *final state* $s' = \{s'_i, i = 1, ..., 6\}$ according to the collision rules described above. A certain *transition probability*

$$A(s \to s') \geq 0 \tag{3.2.7}$$

is assigned to each pair of initial and final state. The transition probabilities satisfy the *normalization*

$$\forall s: \quad \sum_{s'} A(s \to s') = 1 \tag{3.2.8}$$

The only combinations of six real numbers a_i which fulfill the constraints

$$\forall s, s': \quad \sum_i (s'_i - s_i) A(s \to s') a_i = 0 \tag{3.2.9}$$

are linear combinations of 1 (for all i) and c_i, i.e. mass and momentum conservation. This is indeed the case for FHP. The *Zanetti invariants* are *non-local invariants*.

The transition probabilities are invariant with respect to each element of the *isometric group* $\mathcal{G}$:

$$\forall g \in \mathcal{G}, \forall s, s': \quad A(g(s) \to g(s')) = A(s \to s'). \tag{3.2.10}$$

---

[12] Isometries are mappings g which keep the distances $d(\alpha, \beta)$ of arbitrary points α, β invariant: $d(g(\alpha), g(\beta)) = d(\alpha, \beta)$. Rotations and reflections are isometries.

The FHP model fulfills the *detailed balance*

$$\boxed{A(s \to s') = A(s' \to s),}$$ (3.2.11)

i.e. the probabilities for each collision and its inverse collision are equal. For the derivation of the equilibrium distribution (compare Theorem 3.2.1) the weaker condition of *semi-detailed balance* (or *Stueckelberg condition*[13])

$$\boxed{\forall s' : \quad \sum_s A(s \to s') = 1}$$ (3.2.12)

is sufficient.

There are $2^6 = 64$ different states for a node of the FHP-I model. The transition probabilities $A(s \to s')$ form a 64×64 *transition matrix* $A_{ss'}$. The two constraints (3.2.8) and (3.2.12) are equivalent to the statement that the sums of each line or each column of $A_{ss'}$ are equal to 1.

Exercise 3.2.1. (*)
How does the transition matrix $A_{ss'}$ look like? Write down the submatrix which contains the 2- and 3-particle collisions.

Exercise 3.2.2. (**)
Prove: the 2- and 3-particle collisions locally (at a single node) conserve only mass and momentum.

Exercise 3.2.3. (*)
Consider a system with three possible states (a, b, c) and transition probabilities
1.)

$$A(a \to b) = 0.5 \qquad A(b \to a) = 0.5$$
$$A(b \to c) = 0.5 \qquad A(c \to b) = 0.5$$
$$A(c \to a) = 0.5 \qquad A(a \to c) = 0.5$$

or 2.)

$$A(a \to b) = 0.8 \qquad A(b \to a) = 0.2$$
$$A(b \to c) = 0.8 \qquad A(c \to b) = 0.2$$
$$A(c \to a) = 0.8 \qquad A(a \to c) = 0.2$$

---

[13] Stueckelberg (1952) showed that this condition is sufficient to prove the H-theorem.

or 3.)

$$A(a \to b) = 0.8 \qquad A(b \to a) = 0.3$$
$$A(b \to c) = 0.7 \qquad A(c \to b) = 0.4$$
$$A(c \to a) = 0.6 \qquad A(a \to c) = 0.2.$$

Calculate the distribution $(a, b, c)_t$ for large t given the initial value $(a, b, c)_{t=0} = (1, 0, 0)$ for all three cases. In which cases is detailed or semi-detailed balance fulfilled?

3.2.2 Microdynamics of the FHP model

For hydrodynamics certain averaged quantities like mass and momentum density are of interest. To understand the behavior of these averaged quantities a description of lattice-gas cellular automata in terms of statistical mechanics will be formulated. First the analogue to the *Hamilton equations* in classical statistical mechanics will be considered.

Boolean description of the microdynamics. The state of the cells of the lattice-gas cellular automata is described by the *Boolean*[14] arrays $n_i(t, r)$:

$$n_i(t, r) = \begin{cases} 1 & \text{if cell i is occupied} \\ 0 & \text{if cell i is empty} \end{cases}$$

where r and t indicate the discrete points in space and time. The time evolution in terms of Boolean arrays reads:

$$\begin{aligned}
n_i(t+1, r+c_i) = \ & n_i(t, r) \wedge \\
& \{[(n_i \wedge n_{i+1}) \,\&\, (n_{i+1} \wedge n_{i+2}) \,\&\, (n_{i+2} \wedge n_{i+3}) \\
& \&\, (n_{i+3} \wedge n_{i+4}) \,\&\, (n_{i+4} \wedge n_{i+5})] \\
& |\, [n_i \& n_{i+3} \& \sim (n_{i+1} \,|\, n_{i+2} \,|\, n_{i+4} \,|\, n_{i+5})] \\
& |\, [\xi \& n_{i+1} \& n_{i+4} \& \sim (n_i \,|\, n_{i+2} \,|\, n_{i+3} \,|\, n_{i+5})] \\
& |\, [\sim \xi \& n_{i+2} \& n_{i+5} \& \sim (n_i \,|\, n_{i+1} \,|\, n_{i+3} \,|\, n_{i+4})] \}
\end{aligned} \qquad (3.2.13)$$

where the occupation numbers n_i on the right side refer to the input-states at node r and time t. The symbols

---

[14] A short introduction to the Boolean algebra can be found in Appendix 6.1.

$$\begin{aligned}
\& &=& \text{AND} \\
| &=& \text{OR (inclusive or)} \\
\wedge &=& \text{XOR (exclusive or)} \\
\sim &=& \text{NOT.}
\end{aligned} \qquad (3.2.14)$$

have been used. ξ is a Boolean random variable which determines the sense of rotation for the 2-particle collisions (denoted 'xi' in the program code listed below). Eq. (3.2.13) looks rather nasty to people who do not use Boolean algebra every day. The interpretation is however simple. Eq. (3.2.13) states that the state of cell i at node $r + c_i$ at time $t + 1$ is given by $n_i(r, t)$ when no collision happens or (exclusive!) $n_i(r, t)$ is changed by collision before its value is propagated to $r + c_i$. The following collisions are taken into account:

1. A symmetric 3-particle collision happens at node r at time t when the n_j $(j = 1, ..., 6)$ are alternating occupied and empty. When a 3-particle collision happens and n_i is occupied at input cell i will change to the empty state, whereas in a 3-particle collision with n_i empty initially cell i will be occupied after collision.

2. If a 2-particle collision happens where on input cells i and $i + 3$ are occupied (and all other cells empty) n_i will change to 0.

3. If a 2-particle collision happens where on input cells $i + 1$ and $i + 4$ are occupied (and all other cells empty) and the Boolean random variable ξ is 1 such that the rotation of the outgoing particles is in the clockwise direction (otherwise n_i would not change!).

4. If a 2-particle collision happens where on input cells $i + 2$ and $i + 5$ are occupied (and all other cells empty) and the Boolean random variable ξ is 0 such that the rotation of the outgoing particles is in the counter-clockwise direction.

The four different cases correspond to the last five lines on the right hand side of Eq. (3.2.13). n_i is changed if any one of the collisions happens. Therefore the various collision terms are connected by inclusive or. The exclusive or between $n_i(r, t)$ and the collision terms enforces a change when any one of the collisions happens.

The coding of the FHP model is based on the Boolean formulation of the microdynamics. The code in C reads as follows (*nsbit* is the non-solid bit which is set to 1 in the fluid and to 0 inside obstacles; it is used here to suppress collisions inside obstacles):

```
/* loop over all sites */

   for(x=0; x<XMAX; x++) {
   for(y=0; y<YMAX; y++) {
     a = i1[x][y];
     b = i2[x][y];
     c = i3[x][y];
     d = i4[x][y];
     e = i5[x][y];
     f = i6[x][y];

/* two-body collision
   <-> particles in cells a (b,c) and d (e,f)
       no particles in other cells
   <-> db1 (db2,db3) = 1                          */

   db1 = (a&d&~(b|c|e|f));
   db2 = (b&e&~(a|c|d|f));
   db3 = (c&f&~(a|b|d|e));

/* three-body collision <-> 0,1 (bits) alternating
   <-> triple = 1 */

   triple = (a^b)&(b^c)&(c^d)&(d^e)&(e^f);

/* change a and d
   <-> three-body collision triple=1
       or two-body collision db1=1
       or two-body collision db2=1 and xi=1   (- rotation)
       or two-body collision db3=1 and noxi=1 (+ rotation)
   <-> chad=1                                      */

   xi = irn[x][y];        /* random bits */
   noxi = ~xi;

   nsbit = nsb[x][y];    /* non solid bit */

   cha = ((triple|db1|(xi&db2)|(noxi&db3))&nsbit);
   chd = ((triple|db1|(xi&db2)|(noxi&db3))&nsbit);
   chb = ((triple|db2|(xi&db3)|(noxi&db1))&nsbit);
   che = ((triple|db2|(xi&db3)|(noxi&db1))&nsbit);
   chc = ((triple|db3|(xi&db1)|(noxi&db2))&nsbit);
   chf = ((triple|db3|(xi&db1)|(noxi&db2))&nsbit);
```

```
/* change: a = a ^ chad */

    k1[x][y] = i1[x][y]^cha;
    k2[x][y] = i2[x][y]^chb;
    k3[x][y] = i3[x][y]^chc;
    k4[x][y] = i4[x][y]^chd;
    k5[x][y] = i5[x][y]^che;
    k6[x][y] = i6[x][y]^chf;

/* collision finished */
}}
```

Exercise 3.2.4. (**)
Write the analogue of Eq. (3.2.13) for the case where rest particle collisions
(compare Fig. 3.2.2 e) are included.

Arithmetic description of the microdynamics. The transition from the
Boolean to the arithmetic formulation of the microdynamics can be achieved
by formal substitutions (compare Table 3.2.1). However, this procedure is too
laborious. Instead the collision function Δ_i is introduced by

$$\boxed{n_i(t+1, r+c_i) = n_i(t,r) + \Delta_i.}$$ (3.2.15)

The collision function can be constructed according to the following recipe:

- For each collisional configuration, i.e. one that will lead to a change of the
  occupation numbers, a product of the n_i (if $n_i = 1$) respectively $(1 - n_i)$
  (if $n_i = 0$) for $i = 1, ..., b$ (b total number of lattice velocities) is written.
  This product yields 1 if the specific configuration is given and 0 otherwise.

- For nondeterministic collision rules (FHP) the products defined above are
  multiplied by a random variable ξ, $(1 - \xi)$ or 1, respectively.

- The products receive positive or negative sign according to whether n_i
  increases or decreases when changing from input (before collision) to output
  (after collision) state.

- All products are added up.

To give an example let us consider the HPP model. There are only two collisional configurations, namely $(1, 0, 1, 0)$ and $(0, 1, 0, 1)$. The corresponding
products read

$$n_{i+1} n_{i+3}(1 - n_i)(1 - n_{i+2}) \quad \text{and} \quad n_i n_{i+2}(1 - n_{i+1})(1 - n_{i+3})$$

where n_i increase in the former case and decreases in the latter. Thus the
collision function is given by

$$\Delta_i = n_{i+1}n_{i+3}(1 - n_i)(1 - n_{i+2}) - n_i n_{i+2}(1 - n_{i+1})(1 - n_{i+3}).$$

Table 3.2.1. *'Translation' of Boolean expressions into arithmetic formulas.*

a	b	AND	OR	XOR	NOT
		$a \& b$	$a \mid b$	$a \wedge b$	$\sim a$
		$a \cdot b$	$a + b - a \cdot b$	$a + b - 2 \cdot a \cdot b$	$1 - a$
0	0	0	0	0	1
1	0	0	1	1	0
0	1	0	1	1	1
1	1	1	1	0	0

The analogous procedure yields for the FHP-I model

$$
\begin{aligned}
\Delta_i(n) = \ & n_{i+1}n_{i+3}n_{i+5}(1 - n_i)(1 - n_{i+2})(1 - n_{i+4}) \\
& - n_i n_{i+2} n_{i+4}(1 - n_{i+1})(1 - n_{i+3})(1 - n_{i+5}) \\
& + \xi n_{i+1} n_{i+4}(1 - n_i)(1 - n_{i+2})(1 - n_{i+3})(1 - n_{i+5}) \quad (3.2.16) \\
& + (1 - \xi) n_{i+2} n_{i+5}(1 - n_i)(1 - n_{i+1})(1 - n_{i+3})(1 - n_{i+4}) \\
& - n_i n_{i+3}(1 - n_{i+1})(1 - n_{i+2})(1 - n_{i+4})(1 - n_{i+5})
\end{aligned}
$$

where $n = \{n_i, i = 1, ..., 6\}$. The conservation of mass and momentum at each node can be expressed as follows

$$\forall n : \quad \sum_{i=1}^{6} \Delta_i(n) = 0 \tag{3.2.17}$$

and

$$\forall n : \quad \sum_{i=1}^{6} c_i \Delta_i(n) = 0 \tag{3.2.18}$$

These equations imply corresponding conservation laws for the Boolean arrays n_i

$$\sum_i n_i(t + 1, r + c_i) = \sum_i n_i(t, r) \tag{3.2.19}$$

and

$$\sum_i c_i n_i(t + 1, r + c_i) = \sum_i c_i n_i(t, r). \tag{3.2.20}$$

Exercise 3.2.5. (*)
Derive Eq. (3.2.17).

3.2.3 The Liouville equation

Now lattice-gas cellular automata will be considered from the viewpoint of statistical mechanics. In classical statistical mechanics the deterministic description of systems with many degrees of freedom by Hamiltonian equations is abandoned and replaced by a probabilistic approach (*Gibbs' ensemble*: instead of deriving equilibrium values from the long time limit of a single system, one calculates them as mean values over a large hypothetical set of 'equivalent' systems [15]; some authors prefer to avoid the introduction of ensembles: see, for example, Ma, 1993). One can proceed quite similar for lattice-gas cellular automata. The microscopic description of the FHP model encompasses already probabilistic elements (choice of the sense of rotation for 2-particle collisions). To avoid confusion because of these two different types of probabilities the reader should - at least for a while - consider the microdynamics as a deterministic process (like it is indeed for HPP).

Consider a lattice $\mathcal{L}$ of final extend with periodic (cyclic) boundary conditions. The *phase space* Γ (Gibbs) is defined as the *set of all possible states* $s(.)$ of the lattice $\mathcal{L}$. At time $t = 0$ an ensemble of initial states is given with probabilities $P(0, s(.)) \geq 0$ which add up to 1:

$$\sum_{s(.) \in \Gamma} P(0, s(.)) = 1.$$

Each member of the ensemble evolves according to the microdynamics of the lattice gas. This implies the conservation of probabilities

$$P(t + 1, \mathcal{S}s(.)) = P(t, \mathcal{C}^{-1}s(.)) \tag{3.2.21}$$

or

$$P(t + 1, \mathcal{E}s(.)) = P(t, s(.)). \tag{3.2.22}$$

Eq. (3.2.22) is called the *Liouville equation* because of its close analogy to the Liouville equation in classical statistical mechanics (compare page 139). In the case of nondeterministic microdynamics like for FHP the Liouville equation has to be replaced by the more general *Chapman-Kolmogorov equation*

$$P(t + 1, \mathcal{S}s(.)) = \sum_{s(.) \in \Gamma} \prod_{\boldsymbol{r} \in \mathcal{L}} A(s(\boldsymbol{r}) \rightarrow s'(\boldsymbol{r})) P(t, s(.)). \tag{3.2.23}$$

---

[15] Gibbs (1902) writes in his preface: "For some purposes, however, it is desirable to take a broader view of the subject. We may imagine a great number of systems of the same nature, but differing in the configurations and velocities which they have at a given instant, and differing not merely infinitesimally, but it may be so as to embrace every conceivable combination of configuration and velocities. And here we may set the problem, not to follow a particular system through its succession of configurations, but to determine how the whole number of systems will be distributed among the various conceivable configurations and velocities at any required time, when the distribution has been given for some one time."

Exercise 3.2.6. (*)

How many different states are possible a) at a single node and b) on a lattice with N nodes? When does this number exceed 10^{10}?

3.2.4 Mass and momentum density

In the framework of the probabilistic description *ensemble mean values* $q(n(t,...))$ for *observables* q are defined by

$$\langle q(n(t,...)) \rangle := \sum_{s(.) \in \Gamma} q(s(.)) \, P(t, s(.))$$

By far the most important observables are the *mean occupation numbers*

$$N_i(t, r) = \langle n_i(t, r) \rangle$$

which are used to define the mass

$$\rho(t, r) := \sum_i N_i(t, r)$$

and momentum density

$$j(t, r) := \sum_i c_i N_i(t, r).$$

These quantities are defined with respect to nodes and not to cells or area[16]. The density per cell d is calculated by division of ρ by the number of cells per node b ($= 6$ for FHP-I):

$$d = \frac{\rho}{b}. \tag{3.2.24}$$

The flow velocity is defined by the (non-relativistic) relation *momentum density = mass density · velocity*:

$$j(t, r) = \rho(t, r) \, u(t, r).$$

Of course the microscopic conservation equations (3.2.19) and (3.2.20) imply conservation of the averaged quantities

$$\sum_i N_i(t+1, r + c_i) = \sum_i N_i(t, r), \tag{3.2.25}$$

$$\sum_i c_i N_i(t+1, r + c_i) = \sum_i c_i N_i(t, r) \tag{3.2.26}$$

---

[16] The area per node is $\sqrt{3}/2$ (compare Fig. 3.2.1).

3.2.5 Equilibrium mean occupation numbers

After many definitions in the previous subsections now one of the main results of the theoretical analysis of the FHP model will be derived, namely the *equilibrium occupation numbers* N_i^{eq}. Frisch et al. (1987) proved the following theorem which is valid for HPP, FHP and FCHC:

Theorem 3.2.1. *(Frisch et al., 1987)*
The following statements are equivalent:

1. *The* N_i^{eq} *'s are a solution of equation (3.2.23).*

2. *The* N_i^{eq} *'s are a solution of the set of b equations*

$$\forall i = 1, ..., b: \quad \Delta_i(N) := \sum_{ss'}(s_i'-s_i)A(s \to s')\prod_j N_j^{s_j}(1-N_j)^{(1-s_j)} = 0.$$
(3.2.27)

3. *The* N_i^{eq} *'s are given by the Fermi-Dirac distribution*

$$N_i^{eq} = \frac{1}{1+\exp(h+q\cdot c_i)}$$
(3.2.28)

where h is a real number and q is a D-dimensional vector.

Proof. The complete proof is given in Appendix C of Frisch et al. (1987). Here only the step from 2. to 3. will be discussed. The semi-detailed balance condition and the nonexistence of spurious invariants has to be taken into account. In the following the superscript 'eq' will be dropped in order to keep the notation simple.
Define

$$\check{N}_i := \frac{N_i}{1-N_i}$$

and

$$\Pi := \prod_{j=1}^{b}(1-N_j).$$

Eq. (3.2.27) may be written as

$$\frac{\Delta_i}{\Pi} = \sum_{ss'}(s_i'-s_i)A(s \to s')\prod_j \check{N}_j^{s_j} = 0.$$
(3.2.29)

Multiply Eq. (3.2.29) by $\check{N}_i$, sum over i and use

$$\sum_i (s_i'-s_i)\log\check{N}_i = \log\frac{\prod_j \check{N}_j^{s_j'}}{\prod_j \check{N}_j^{s_j}}$$

to obtain

$$\sum_{ss'} A(s \rightarrow s') \log \left(\frac{\prod_j \check{N}_j^{s'_j}}{\prod_j \check{N}_j^{s_j}} \right) \prod_j \check{N}_j^{s_j} = 0. \qquad (3.2.30)$$

Semi-detailed balance

$$\sum_s A(s \rightarrow s') = \sum_{s'} A(s \rightarrow s') = 1$$

implies that

$$\sum_{ss'} A(s \rightarrow s') \left(\prod_j \check{N}_j^{s_j} - \prod_j \check{N}_j^{s'_j} \right) = 0. \qquad (3.2.31)$$

Combining Eqs. (3.2.30) and (3.2.31), one obtains

$$\sum_{ss'} A(s \rightarrow s') \left[\log \left(\frac{\prod_j \check{N}_j^{s'_j}}{\prod_j \check{N}_j^{s_j}} \right) \prod_j \check{N}_j^{s_j} + \prod_j \check{N}_j^{s_j} - \prod_j \check{N}_j^{s'_j} \right] = 0. \qquad (3.2.32)$$

The relation ($x > 0$, $y > 0$)

$$y \log \frac{x}{y} + y - x = - \int_x^y \log \frac{t}{y} dt \le 0 \qquad (3.2.33)$$

where equality being achieved only when $x = y$ will be exploited. The left hand side of (3.2.32) is a linear combination of expressions of the form (3.2.33) with nonnegative weights $A(s \rightarrow s')$. For it to vanish, one must have

$$\prod_j \check{N}_j^{s_j} = \prod_j \check{N}_j^{s'_j} \quad \text{whenever} \quad A(s \rightarrow s') \neq 0.$$

This is equivalent to

$$\forall s, s' : \quad \sum_i \log(\check{N}_i)(s'_i - s_i) A(s \rightarrow s') = 0. \qquad (3.2.34)$$

Eq. (3.2.34) means that $\log(\check{N}_i)$ is a collision invariant. Now assuming that only mass and momentum are conserved and no spurious invariants exist, one concludes that

$$\log(\check{N}_i) = -(h + \mathbf{q} \cdot \mathbf{c}_i),$$

which is the most general collision invariant (a linear combination of the mass invariant and of the D momentum invariants). Reverting to the mean populations $N_i = \check{N}_i/(1 + \check{N}_i)$, one obtains (3.2.28). **q.e.d.**

Calculation of the Lagrange multipliers at small flow speeds. The determination of the Lagrange multipliers[17] h and q is constrained by the conserved quantities

$$\rho = \sum_i N_i = \sum_i \frac{1}{1 + \exp(h + q \cdot c_i)} \qquad (3.2.35)$$

and

$$\rho u = \sum_i N_i c_i = \sum_i \frac{c_i}{1 + \exp(h + q \cdot c_i)}. \qquad (3.2.36)$$

These equations apply also for HPP where the problem of the determination of the Lagrange multipliers can be reduced to a cubic equation (Hardy et al., 1973). In contrast to HPP, for the FHP model explicit solutions are known only in a few special cases.

The Lagrange multipliers can be calculated by an expansion for small *Mach numbers* $M_a := U/c_s$, i.e. for speeds $U = |u|$ well below the sound speed c_s. The somewhat lengthy calculations in Appendix 6.2 apply to FHP and HPP.

The equilibrium distributions for FHP-I read

$$\boxed{N_i^{eq}(\rho, u) = \frac{\rho}{6} + \frac{\rho}{3} c_i \cdot u + \rho G(\rho) Q_{i\alpha\beta} u_\alpha u_\beta + \mathcal{O}\left(u^3\right)} \qquad (3.2.37)$$

with

$$G(\rho) = \frac{1}{3} \frac{6 - 2\rho}{6 - \rho} \qquad (3.2.38)$$

and

$$Q_{i\alpha\beta} = c_{i\alpha} c_{i\beta} - \frac{1}{2} \delta_{\alpha\beta}. \qquad (3.2.39)$$

The term quadratic in u will lead to the nonlinear advection term in the Navier-Stokes equation. This is the reason to expand the N_i^{eq} up to second order.

Frisch et al. (1987) have shown that the generalization of (3.2.37) to (3.2.39), namely

$$N_i^{eq}(\rho, u) = \frac{\rho}{b} + \frac{\rho D}{c^2 b} c_{i\alpha} u_{i\alpha} + \rho G(\rho) Q_{i\alpha\beta} u_\alpha u_\beta + \mathcal{O}\left(u^3\right) \qquad (3.2.40)$$

with

$$G(\rho) = \frac{D^2}{2c^4 b} \frac{b - 2\rho}{b - \rho} \qquad (3.2.41)$$

and

$$Q_{i\alpha\beta} = c_{i\alpha} c_{i\beta} - \frac{c^2}{D} \delta_{\alpha\beta}. \qquad (3.2.42)$$

(b number of cells per node, D dimension) is valid for HPP, FHP-I, FHP-II, FHP-III and FCHC.

---

[17] The Fermi-Dirac distribution (3.2.28) could be derived from a *maximum entropy principle* whereby the constraints of mass and momentum conservation are coupled to the entropy (Shannon) by Lagrange multipliers like h and q. This method will be discussed in some detail in Section 4.3 and applied in Section 5.2.

Exercise 3.2.7. (*)
Specify the equilibrium distribution for HPP and compare it with the linear
distribution (3.1.13) which was used for initialization in Section 3.1.

3.2.6 Derivation of the macroscopic equations: multi-scale analysis

The derivation of macroscopic equations by *multi-scale analysis* is one of
the most demanding topics in the whole book. In the current section the
expansion will be followed up to first order only. The calculation of all second
order terms is too involved to be shown here in detail. The multi-scale analysis
will be discussed again in a special section devoted to the Chapman-Enskog
expansion (Section 4.2) and will be applied up to second order in Section 5.2
in the context of lattice Boltzmann models.

In the former subsection the distribution functions for a *global* (homogeneous)
equilibrium were derived. The interesting aspects of fluid flows and of nature
in general lie, however, in its variations in space and time. One can think of
the 'real world' as a *patchwork of thermodynamic equilibria* whose parameters
like mass, momentum or energy[18] density show slow changes in space, such
that every point can be characterized by the *local* values of mass, momentum
and energy density.

For FHP equilibrium mean occupation numbers N_i^{eq} have been derived which
depend continuously on tunable parameters, namely the mean values of the
conserved quantities mass and momentum density. At the beginning of a
numerical simulation a distribution of mass and momentum density will be
initialized which varies on large (compared to the lattice unit[19]) spatial scales
ϵ^{-1} (measured in lattice units; ϵ is a small number). In the course of time
three phenomena can be distinguished with respect to their characteristic
time scales:

1. *Relaxation toward local equilibrium* with time scale ϵ^0: very fast. Few
   collision are necessary to reach local equilibrium (compare Fig. 3.2.6).
   Note that the number of collisions per time interval is a function of mass
   density. Thus the characteristic time scale ϵ^0 is large at low and high
   mass densities where collisions are rare.

2. Sound waves (perturbations of mass density) and advection with time
   scale ϵ^{-1}: fast, but slower than relaxation toward local equilibrium.

3. Diffusion with time scale ϵ^{-2}: distinctly slower than sound waves and
   advection.

---

[18] Energy does not play a role for FHP and many other LGCA because it does
not exist as an independent quantity or because it is not conserved by certain
collisions whereas so-called thermal LGCA include an energy equation (compare
Section 3.7).

[19] Lattice unit = distance between neighboring nodes = 1 for FHP

Let us consider, for example, spatial variations on the scale of $\epsilon^{-1} = 100$ lattice units which may be due to an obstacle in a flow (compare the von Karman vortex street shown in Fig. 3.6.5). The relaxation toward local equilibrium proceeds on a time-scale of order $\epsilon^0 = 1$, i.e. in a few time steps[20]. Sound waves[21] and advection show time scales of the order of $\epsilon^{-1} = 100$ whereas diffusion with time-scales of $\epsilon^{-2} = 10\,000$ is much slower.

The microdynamics of the lattice-gas cellular automata contain all of these phenomena. We will now apply the so-called *multi-scale technique* to 'pick out' the processes of interest here, namely the hydrodynamic modes. Corresponding to the differentiation given above one introduces three time variables:

$$t_\star \qquad \text{(discrete)}$$
$$t_1 = \epsilon t_\star$$
$$t_2 = \epsilon^2 t_\star$$

where the last two scales will be considered as continuous[22] variables. With respect to space only two scales have to be distinguished because sound waves and advection as well as diffusion act on similar spatial scales:

$$r_\star \qquad \text{(discrete)}$$
$$r_1 = \epsilon r_\star \quad \text{(continuous)}$$

$N_i^{(0)}$ will refer to mean occupation numbers for given local values of ρ and u. The $N_i^{(0)}$ are given by the *global* form of the equilibrium distributions (3.2.37) but where ρ and u are the *local* values of mass density and velocity, i.e. $N_i^{(0)}(r, t) \equiv N_i^{eq}(\rho, u)$. The actual occupation numbers $N_i(t, r)$ are close to the equilibrium values and therefore can be expanded about $N_i^{(0)}$:

$$N_i = N_i^{(0)}(t, r) + \epsilon N_i^{(1)}(t, r) + \mathcal{O}\left(\epsilon^2\right) \qquad (3.2.43)$$

Terms of higher than linear order in ϵ will be neglected. The linear corrections do not contribute to the local values of mass and momentum density:

$$\sum_i N_i^{(1)}(t, r) = 0 \quad \text{and} \quad \sum_i c_i N_i^{(1)}(t, r) = 0 \qquad (3.2.44)$$

which follows from

---

[20] At reasonable mass densities, because otherwise there are not enough collisions.
[21] Impressive sound waves are created, for example, when an obstacle like the plate in Fig. 3.6.5 is suddenly put into an initially homogeneous flow.
[22] Over longer time scales a thing or two are smoothed out.

$$\rho = \sum_i N_i(t, r) = \sum_i N_i^{(0)}(t, r) \quad \text{and} \qquad (3.2.45)$$

$$j = \sum_i c_i N_i(t, r) = \sum_i c_i N_i^{(0)}(t, r). \qquad (3.2.46)$$

In the *Chapman-Enskog expansion* of the Boltzmann equation (Chapman, 1916, 1918; Enskog, 1917; Chapman and Cowling, 1970; Cercignani, 1990) ϵ is the *Knudsen number* K_n, i.e. the ratio between the mean free path length l and the characteristic length scale of the system L which can be the diameter of an obstacle (for example: flow past a sphere) or the size of the whole domain. The hydrodynamic (continuous) regime is characterized by small Knudsen numbers whereas finite size effects play a role at Knudsen number of order 1 or higher (Knudsen flows).

The microscopic conservation laws (3.2.25) and (3.2.26) are the starting point for the multi-scale analysis. The mean populations after collision and propagation are expanded up to second order in ϵ around its values before the collision step:

$$\left\{ \begin{matrix} 1 \\ c_i \end{matrix} \right\} N_i(t+1, r + c_i) = \left\{ \begin{matrix} 1 \\ c_i \end{matrix} \right\} [\, N_i(t, r) + \partial_t N_i + c_{i\alpha}\partial_{x_\alpha} N_i$$
$$+ \frac{1}{2}\partial_t\partial_t N_i + \frac{1}{2}c_{i\alpha}c_{i\beta}\partial_{x_\alpha}\partial_{x_\beta} N_i \qquad (3.2.47)$$
$$+ c_{i\alpha}\partial_t\partial_{x_\alpha} N_i + \mathcal{O}\left(\partial^3 N_i\right) \,]$$

In what follows the fast (local) relaxation processes will be neglected in the theoretical description because one is only interested in the hydrodynamic behavior of the lattice-gas cellular automata. The derivations in time and space are substituted in Eq. (3.2.47) according to the scalings given above:

$$\partial_t \quad \longrightarrow \quad \epsilon\partial_t^{(1)} + \epsilon^2\partial_t^{(2)} \qquad (3.2.48)$$
$$\partial_{x_\alpha} \quad \longrightarrow \quad \epsilon\partial_{x_\alpha}^{(1)}. \qquad (3.2.49)$$

Insertion of Eqs. (3.2.43), (3.2.47), (3.2.48) and (3.2.49) into Eqs. (3.2.25) and (3.2.26) leads to

$$\sum_i \left\{ \begin{matrix} 1 \\ c_i \end{matrix} \right\} [N_i(t+1, r + c_i) - N_i(t, r)]$$

$$= \sum_i \left\{ \begin{matrix} 1 \\ c_i \end{matrix} \right\} \left[\epsilon\partial_t^{(1)} N_i^{(0)} + \epsilon^2\partial_t^{(1)} N_i^{(1)} + \epsilon^2\partial_t^{(2)} N_i^{(0)} + \epsilon^3\partial_t^{(2)} N_i^{(1)} \right]$$

$$+\epsilon c_{i\alpha}\partial_{x_\alpha}^{(1)}N_i^{(0)} + \epsilon^2 c_{i\alpha}\partial_{x_\alpha}^{(1)}N_i^{(1)} + \frac{1}{2}\epsilon^2\partial_t^{(1)}\partial_t^{(1)}N_i^{(0)} + \frac{1}{2}\epsilon^3\partial_t^{(1)}\partial_t^{(1)}N_i^{(1)}$$

$$+\epsilon^3\partial_t^{(1)}\partial_t^{(2)}N_i^{(0)} + \epsilon^4\partial_t^{(1)}\partial_t^{(2)}N_i^{(1)} + \epsilon^4\partial_t^{(2)}\partial_t^{(2)}N_i^{(0)} + \epsilon^5\partial_t^{(2)}\partial_t^{(2)}N_i^{(1)}$$

$$+\frac{1}{2}\epsilon^2 c_{i\alpha}c_{i\beta}\partial_{x_\alpha}^{(1)}\partial_{x_\beta}^{(1)}N_i^{(0)} + \frac{1}{2}\epsilon^3 c_{i\alpha}c_{i\beta}\partial_{x_\alpha}^{(1)}\partial_{x_\beta}^{(1)}N_i^{(1)} + \epsilon^2 c_{i\alpha}\partial_t^{(1)}\partial_{x_\alpha}^{(1)}N_i^{(0)}$$

$$+\epsilon^3 c_{i\alpha}\partial_t^{(1)}\partial_{x_\alpha}^{(1)}N_i^{(1)} + \epsilon^3 c_{i\alpha}\partial_t^{(2)}\partial_{x_\alpha}^{(1)}N_i^{(0)} + \epsilon^4 c_{i\alpha}\partial_t^{(2)}\partial_{x_\alpha}^{(1)}N_i^{(1)}\Big]$$

$$= 0.$$

To first order in ϵ one obtains

$$\partial_t^{(1)}\sum_i N_i^{(0)} + \partial_{x_\beta}^{(1)}\sum_i c_{i\beta}N_i^{(0)} = 0 \tag{3.2.50}$$

and

$$\partial_t^{(1)}\sum_i c_{i\alpha}N_i^{(0)} + \partial_{x_\beta}^{(1)}\sum_i c_{i\alpha}c_{i\beta}N_i^{(0)} = 0 \tag{3.2.51}$$

or

$$\partial_t^{(1)}\rho + \partial_{x_\beta}^{(1)}(\rho u_\beta) = 0 \quad \text{(continuity equation)} \tag{3.2.52}$$

$$\partial_t^{(1)}(\rho u_\alpha) + \partial_{x_\beta}^{(1)}P_{\alpha\beta}^{(0)} = 0 \tag{3.2.53}$$

where

$$P_{\alpha\beta}^{(0)} \equiv \sum_i c_{i\alpha}c_{i\beta}N_i^{eq}$$

$$= \rho\delta_{\alpha\beta} + \underbrace{\rho G(\rho)T_{\alpha\beta\gamma\delta}^{(MA)}u_\gamma u_\delta}_{\longrightarrow \text{ advection term}} + \mathcal{O}(u^4) \tag{3.2.54}$$

is the momentum flux tensor in first approximation. The momentum advection tensor[23] $T^{(MA)}$ is a tensor of 4th rank

$$T_{\alpha\beta\gamma\delta}^{(MA)} = \sum_i c_{i\alpha}c_{i\beta}Q_{i\gamma\delta}. \tag{3.2.55}$$

It is isotropic and given by (compare Section 3.3, Eq. (3.3.10)):

$$T_{\alpha\beta\gamma\delta}^{(MA)} = \sum_i c_{i\alpha}c_{i\beta}Q_{i\gamma\delta} = \sum_i c_{i\alpha}c_{i\beta}\left(c_{i\gamma}c_{i\delta} - \frac{1}{2}\delta_{\gamma\delta}\right)$$

$$= \frac{3}{4}\left(\delta_{\alpha\gamma}\delta_{\beta\delta} + \delta_{\alpha\delta}\delta_{\beta\gamma} - \delta_{\alpha\beta}\delta_{\gamma\delta}\right).$$

---

[23] The name is derived from the fact that this tensor is part of the nonlinear advection term. It occurs, however, also in the dissipative terms of the Navier-Stokes equation.

Accordingly the components of the momentum flux tensor in first approximation read

$$
\begin{aligned}
P_{xx}^{(0)} &= \rho G(\rho)\left(u^2 - v^2\right) + \frac{\rho}{2} \\
P_{yy}^{(0)} &= \rho G(\rho)\left(v^2 - u^2\right) + \frac{\rho}{2} \\
P_{xy}^{(0)} &= P_{yx}^{(0)} = \rho G(\rho) 2uv
\end{aligned}
\qquad (3.2.56)
$$

whereas the momentum flux tensor in the 'real world' (i.e. in the Navier-Stokes equation) is

$$
\begin{aligned}
P_{xx} &= \rho u^2 + p \\
P_{yy} &= \rho v^2 + p \\
P_{xy} &= P_{yx} = \rho uv.
\end{aligned}
\qquad (3.2.57)
$$

Identification of $\frac{\rho}{2}(1 - g(\rho)u^2)$ with the pressure p leads to

$$
\begin{aligned}
P_{xx}^{(0)} &= \rho g(\rho)u^2 + p \\
P_{yy}^{(0)} &= \rho g(\rho)v^2 + p \\
P_{xy}^{(0)} &= P_{yx}^{(0)} = \rho g(\rho)uv
\end{aligned}
\qquad (3.2.58)
$$

which looks similar to the momentum flux tensor (3.2.57) except for the factor $g(\rho) = G(\rho)/2$. For small values of u^2 the pressure is given by the 'isothermal' relation

$$
p = \frac{\rho}{2} = c_s^2 \rho \qquad (3.2.59)
$$

with the sound speed $c_s = 1/\sqrt{2}$. It can be shown that invariance under *Galilei transformations* constrains the *g-factor* to be equal to 1 (compare Exercise 3.2.8). Here

$$
g(\rho) = \frac{3 - \rho}{6 - \rho} \qquad (3.2.60)
$$

is always smaller than 1 (actually smaller than 1/2). Similar expressions apply to other lattice models. Thus this *g*-factor *breaks* the Galilean invariance. The deviation of *g* from 1 is caused by the underlying lattice which is only invariant under certain discrete translations but not under arbitrary Galilei transformations. It will be shown later that this 'disease' can be cured when other distribution functions (Boltzmann instead of Fermi-Dirac) are applied.

This is the case in lattice Boltzmann models. The occurance of Fermi-Dirac distributions in LGCA is a consequence of the exclusion principle. Therefore if one sticks to this essential feature of any LGCA the g-disease can be treated only symptomatically[24], namely by a *rescaling of time*

$$t \quad \longrightarrow \quad \frac{t}{g(\rho)}. \tag{3.2.61}$$

The equation for the incompressible regime can be derived from the compressible equation by ignoring all density variations except in the pressure term[25]. Setting ρ to the constant and uniform value ρ_0 and applying the rescaling of time leads to

$$\rho_0 g(\rho_0)\frac{\partial u}{\partial t} + g(\rho_0)(u\nabla)u = -\nabla\left(\frac{\rho}{2} - \frac{\rho_0}{2}g(\rho_0)u^2\right)$$

or

$$\frac{\partial u}{\partial t} + (u\nabla)u = -\nabla P \tag{3.2.62}$$

with the kinematic pressure

$$P = \left(\frac{\rho}{2\rho_0 g(\rho_0)} - u^2\right) \tag{3.2.63}$$

Eq. (3.2.62) is the Euler equation (Navier-Stokes without dissipation) of the FHP model. Eq. (3.2.52) is the continuity equation for the mass density ρ. It will not change when terms of order ϵ^2 are included.

The terms of order ϵ^2 are calculated by Frisch et al. (1987) and Hénon (1987b). Adding up terms of order ϵ and ϵ^2 while neglecting terms of order ϵu^3, $\epsilon^2 u^2$ and $\epsilon^3 u$ leads in the same incompressible limit as discussed before to the Navier-Stokes equation

$$\boxed{\begin{aligned} \nabla \cdot u &= 0 \\ \partial_t u + (u\nabla)u &= -\nabla P + \nu\nabla^2 u \end{aligned}} \tag{3.2.64}$$

where ν is the (scaled) kinematic shear viscosity ($\nu = \nu^{(u)}/g(\rho_0)$; the unscaled shear viscosity $\nu^{(u)}$ for different FHP models is given in Table 3.2.2; compare also Fig. 3.2.4). Thus we have attained the object of our desire: the incompressible Navier-Stokes equation!

Please note that the FHP models do not possess an energy equation. In the FHP-I model mass and (kinetic) energy conservation are essentially identical. In models with rest particles, some collisions do not respect conservation of

Fig. 3.2.3. *The Reynolds coefficient R_* as a function of the density per cell d for FHP-I, FHP-II and FHP-III. FHP-I: 2-particle and symmetric 3-particle collisions; FHP-II: additional collisions involving rest particles and 2-particle collisions with spectator; FHP-III: additional 4-particle collisions (collision saturated model). The Reynolds coefficient is defined as $R_* := \dfrac{R_e}{L \cdot M_a} = \dfrac{c_s g(d)}{\nu(d)}$; $R_*^{\max} = \max_d (R_*) = R_*(d_{max})$; d is the density per cell (Eq. 3.2.24), c_s sound speed, $g(d)$ density dependent g-factor, and $\nu^{(u)}$ (unscaled) kinematic shear viscosity.*

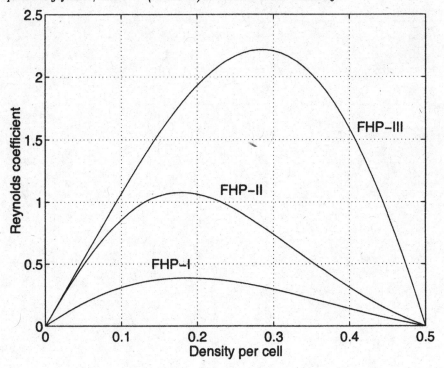

(kinetic) energy. lattice-gas cellular automata with energy equations will be discussed in Section 3.7.

The symptomatic treatment of the g-disease does not solve all problems. D'Humières et al. (1987) have shown that vorticity is advected at speed $g(d)u \neq u$. In order to fix this problem they have proposed a model with 8 bits over the FHP lattice (see Subsection 3.2.10 for details).

[25] For some fine points see Majda, 1984.

Table 3.2.2. *Analytical values for three different FHP models.* $\nu^{(u)}$ *(unscaled) kinematic shear viscosity;* $\eta^{(u)}$ *(unscaled) kinematic bulk (compressional) viscosity (taken from Frisch et al., 1987). d is the density per cell (Eq. 3.2.24).*

	FHP-I	FHP-II	FHP-III
Number of cells	6	7	7
c_s	$1/\sqrt{2}$	$\sqrt{3/7}$	$\sqrt{3/7}$
$g(d)$	$\dfrac{1}{2}\dfrac{1-2d}{1-d}$	$\dfrac{7}{12}\dfrac{1-2d}{1-d}$	$\dfrac{7}{12}\dfrac{1-2d}{1-d}$
$\nu^{(u)}$	$\dfrac{1}{12}\dfrac{1}{d(1-d)^3}-\dfrac{1}{8}$	$\dfrac{1}{28}\dfrac{1-4d/7}{d(1-d)^3}-\dfrac{1}{8}$	$\dfrac{1}{28}\dfrac{1-8d(1-d)/7}{d(1-d)}-\dfrac{1}{8}$
$\eta^{(u)}$	0	$\dfrac{1}{98}\dfrac{1}{d(1-d)^4}-\dfrac{1}{28}$	$\dfrac{1}{98}\dfrac{1-2d(1-d)}{d(1-d)}-\dfrac{1}{28}$
$R_*^{\max}$	0.387	1.08	2.22
d_{max}	0.187	0.179	0.285

Fig. 3.2.4. *The g-factor, unscaled kinematic bulk viscosity ($\xi^{(u)}$), and the unscaled ($\nu^{(u)}$) and scaled (ν) kinematic shear viscosity as functions of the density per cell (d) for the models FHP-I, FHP-II and FHP-III (compare Table 3.2.2). The bulk viscosity vanishes in FHP-I.*

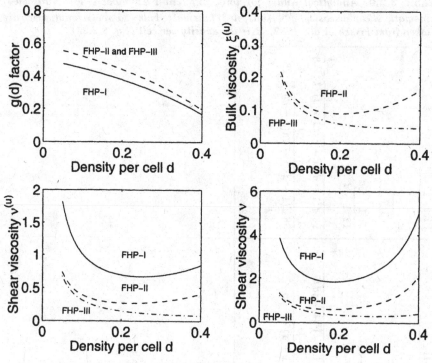

Exercise 3.2.8. (**)
Show that the generalized (g-factor) substantial derivative

$$\frac{\partial \mathbf{u}}{\partial t} + g\,(\mathbf{u}\boldsymbol{\nabla})\,\mathbf{u} \tag{3.2.65}$$

is invariant under Galilei transformations if and only if $g \equiv 1$.
Hint: it will be sufficient to consider the special Galilei transformation

$$
\begin{aligned}
x' &= x + ct; \quad c = const \\
y' &= y \\
\\
z' &= z \\
t' &= t
\end{aligned}
\tag{3.2.66}
$$

Exercise 3.2.9. (**)

Calculate the components of the momentum flux tensor $P_{\alpha\beta}^{(0)}$ for HPP from the formula

$$\rho G(\rho) T_{\alpha\beta\gamma\delta}^{(MA)} + \frac{c^2}{D}\rho\delta_{\alpha\beta}. \tag{3.2.67}$$

What's wrong with the advection term?

3.2.7 Boundary conditions

The coding of boundary conditions (BC) is an essential part of the LGCA (and any other numerical) method. There are at least five different types of BC:

1. *Periodic* BC are often used even if it is not realistic because they are so easy to code.

2. *Inflow* BC (example: channel flow).

3. *Outflow* BC (example: channel flow) can be very difficult to deal with, especially when waves try to leave the model domain (compare, for example, Orlanski, 1976, Røed and Smedstad, 1984, and Stevens, 1991).

4. *No-slip* BC ($u = 0$) apply to solid boundaries (walls, obstacles).

5. *Slip* BC, i.e. the velocity component normal to the boundary and the normal derivative of the tangential component vanish ($u_n = 0$ and $\partial u_t/\partial n = 0$), apply to solid boundaries where the frictional force adjacent to the wall is not resolved.

Even when constraints are formulated only for the velocity one usually also requires conservation of mass[26]. Whereas in channel flows a small violation of mass conservation could be tolerable, because each fluid element leaves the domain after some time anyway, such a violation is not acceptable for flows in closed domains where a small but steady leakage, for example, would lead to an empty basin after a while.

The coding of the first two types of BC is obvious not only for FHP but for any kind of LGCA or LBM and therefore needs no further comment. Since the beginning of simulations with FHP the following heuristic procedures for the implementation of no-slip and slip conditions are used (d'Humières and Lallemande, 1987):

– No-slip: Collision on boundary points are skipped. Instead the incoming particles are turned around by 180°. In the next propagation step they will

---

[26] However, momentum is most often not conserved at the boundaries.

leave the node in their former incoming direction. The mean value over in-state and out-state yields $u = 0$. This flipping of the incoming particles is also called *bounce-back rule*. The kinetic equation for this implementation of the no-slip condition reads

$$n_i(x_b + c_i, t + 1) = n_{i+3}(x_b, t)$$

where x_b is a boundary point and as usual the indices are understood mod-ulo 6.

This *node-oriented implementation* of the no-slip BC has the advantage that it is independent of the orientation of the wall (for alternative imple-mentations see Rem and Somers, 1989, Cornubert et al., 1991, and Ziegler, 1993).

– Slip: Collisions on boundary points are skipped. Instead the incoming par-ticles are reflected like a light ray (*specular reflection*) at the wall. The mean value over in-state and out-state yields $u_n = 0$ and the tangential momentum of the particles is conserved. The kinetic equation for this im-plementation of the no-slip condition reads

$$n_i(x_b + c_i, t + 1) = n_{i-3}(x_b, t).$$

Only several years later Cornubert et al. (1991) have investigated discrepan-cies of the distribution functions in the boundary layer due to these kind of im-plementations of boundary conditions. For FHP they found anisotropic Knud-sen layers adjacent to obstacles. The effective boundary of obstacles is not identical with the node locations but lies somewhat outside the outer nodes of the obstacle. The precise location depends on the direction (anisotropic) and is smaller than the distance to the next neighbor node in the fluid region. Especially for small obstacles this effect has to be taken into account.

3.2.8 Inclusion of body forces

In principle body forces, i.e. a change of momentum, can be applied on the macroscopic or microscopic level. The macroscopic method, however, requires at each time step the calculation of mean values (coarse graining), change of the momentum j and re-initilization. Of course this procedure is computa-tionally much too demanding. Thus only a microscopic method is feasable.

Body forces F which may vary in space and time but are independent of the flow velocity u can be realized by flipping particles with velocity $-c_i$ into particles with velocity c_i (for forces parallel to c_i). The probability of this flipping has to be proportional to the magnitude of F.

The following results are from Appendix D in Frisch et al. (1987). Boolean transition variables $\xi'_{ss'}$ are defined such that their mean values

$$\langle \xi'_{ss'} \rangle = B(s \rightarrow s') \tag{3.2.68}$$

are a set of transition probabilities associated to the body-force. The transition probabilities satisfy normalization

$$\sum_{s'} B(s \to s') = 1 \tag{3.2.69}$$

and mass conservation

$$\sum_i (s'_i - s_i) B(s \to s') = 0 \quad \forall s, s', \tag{3.2.70}$$

but *do not* satisfy momentum conservation (which is desired of course!), semi-detailed balance and $\mathcal{G}$-invariance. In case I (body-force f independent of velocity) they are further constrained by

$$f = \sum_{s,s',i} c_i (s'_i - s_i) B(s \to s') \left(\frac{d}{1-d}\right)^p (1-d)^b \tag{3.2.71}$$

where $p = \sum_j s_j$ and b is the number of cells per node. If f is space and/or time dependent, so are the $B(s \to s')$'s.

In case of a force linear in velocity

$$f_\alpha = C_{\alpha\beta} U_\beta \tag{3.2.72}$$

the $B(s \to s')$'s are constrained by

$$0 = \sum_{s,s',i} c_i (s'_i - s_i) B(s \to s') \left(\frac{d}{1-d}\right)^p (1-d)^b, \quad p = \sum_j s_j, \tag{3.2.73}$$

and

$$C_{\alpha\beta} = \frac{D}{c^2} (1-d)^{b-1} \sum_{s,s',i} c_{i\alpha} (s'_i - s_i) B(s \to s') \left(\frac{d}{1-d}\right)^p \sum_j s_j c_{j\beta} \tag{3.2.74}$$

where $p = \sum_j s_j$.

In order to illustrate the method let us consider the simplest example, namely time-independent homogeneous forcing in the direction of a particular lattice velocity, say in x-direction. We can flip particles with velocity $-c_6 = c_3 = (-1,0)$ into particles with velocity $c_6 = (1,0)$ whenever this is possible while leaving all other particles unchanged (compare Fig. 3.2.5). Already after one time step the domain mean x-velocity, $u_x(t)$, increased from zero (its initial value) to approximately 0.2. This extreme acceleration leads to high Mach numbers after a few time steps. Thus in simulations of incompressible flow problems the flipping rate has to be lowered.

Fig. 3.2.5. *Inclusion of body-forces: The plot shows the increase of domain mean x-velocity, $u_x(t)$, due to flipping of particles with velocity $-c_6 = c_3 = (-1, 0)$ into particles with velocity $c_6 = (1, 0)$ whenever this is possible (FHP-I, d = 0.3).*

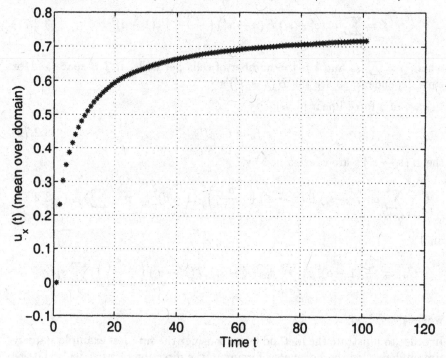

3.2.9 Numerical experiments with FHP

Here the results of some numerical calculations will be discussed. The code written in C is accessible (see the web address given in the Preface).

1. Relaxation toward equilibrium (compare program 'exper1.c'): The FHP-I model is initialized with a distribution which is spatially homogenous but far from equilibrium. At a density $\rho \approx 2$ enough collisions occur in order to drive the occupation numbers toward their equilibrium values (compare Fig. 3.2.6).

Fig. 3.2.6. *Relaxation toward equilibrium for FHP-I: The model is initialized with a distribution which is spatially homogenous but far from equilibrium. The domain size is 320 times 320 nodes. For simplicity periodic boundary conditions are applied. At a density $\rho \approx 2$ enough collisions occur in order to drive the occupation numbers toward their equilibrium values (dashed lines) which are almost reached after 10 time steps. The mean values over the time levels 30 to 60 compares very well with the theoretical values according to equilibrium distribution (3.2.37).*

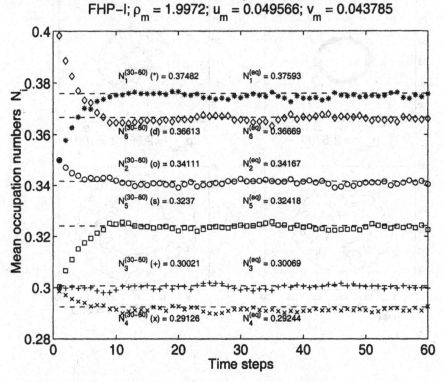

FHP–I; $\rho_m = 1.9972$; $u_m = 0.049566$; $v_m = 0.043785$

2. Propagation of *sound waves*: Here sound waves are excited by a density perturbation (compare Fig. 3.2.7 upper left). The waves propage isotropically with a speed of $c_s = 1/\sqrt{2}$.

Fig. 3.2.7. *Propagation of sound waves: The FHP-I model is initialized with vanishing velocity and constant density (≈ 2) except for a positive radial symmetric density perturbation near the center of a domain with 960 times 960 nodes. The plots show the mean densities calculated over subdomains (macrocells) of 32 times 32 nodes at four different time levels. The values of the contour lines are always $2.1, 2.2, 2.3, 2.4, 2.5$. The propagation of the density perturbation is isotropic and the speed of propagation is consistent with the sound speed $c_s = 1/\sqrt{2}$ (in 300 time steps the sound will propagate over a distance of $300/32/\sqrt{2} \approx 13$ sidelengths of macrocells).*

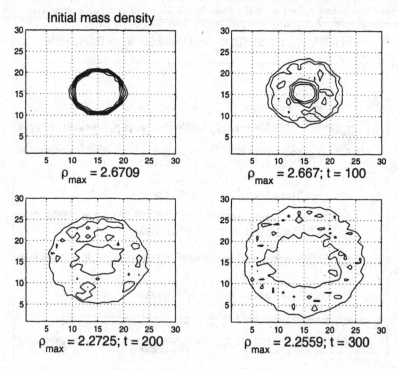

3. Flow past an obstacle: Our next goal is the simulation of the flow past an obstacle. Wether the flow is laminar or turbulent depends on the Reynolds number R_e which is defined by

$$R_e := \frac{U \cdot L}{\nu} \qquad (3.2.75)$$

where U is a characteristic flow speed (usually the flow speed far upstream of the obstacle), L is a characteristic spatial scale of the obstacle (for an obstacle in the form of a circular cylinder, for example, L is the radius or diameter of the cross-section), and ν is the kinematic shear viscosity. From experiments it is known that eddies are formed and shedded when the Reynolds number becomes larger than about 50. A so-called von Karman vortex street will build up (compare the beautiful pictures in the book of van Dyke, 1982).

How can one simulate a flow with $R_e \approx 90$ with FHP-II? The flow speed has to be small compared to the sound speed $c_s^{(\text{FHP-II})} = \sqrt{3/7} \approx 0.65$. So $u = 0.2$ is a good value. The density per cell d should be below 0.5 (the scaling factor $g(d)$ vanishes at $d = 0.5$) and is chosen here as $d = 2/7$ which results in a density per node $\rho = 2$. The scaled kinematic viscosity at this density is approximately 0.8. The only free parameter is the size of the obstacle L which can be calculated from

$$L = \frac{R_e \cdot \nu}{u} = \frac{88 \cdot 0.88}{0.2} = 388 \qquad (3.2.76)$$

(see results in Fig. (3.2.8)).

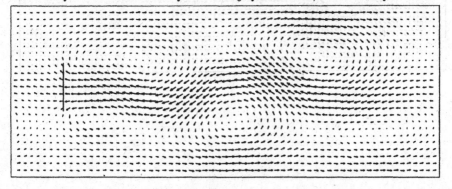

Fig. 3.2.8. *Flow past a plate. The FHP-II model is initialized with a homogenous flow in x-direction with speed $u = 0.2$ (in lattice units). The lattice encompasses 4096 times 1792 nodes. The width of the plate is 388. At a density per node $\rho = 2.1$ the (scaled) kinematic viscosity is 0.88. Thus the Reynolds number is 88. The figure shows the flow minus the mean flow velocity after $t = 260,000$ time steps.*

Please note that the type of boundary conditions are essential. If one applies periodic boundary conditions the simulation addresses flow past a periodic array of cylinders. In contrast to flow past a single cylinder the flow may be steady even at a Reynolds number of 100 (Gallivan et al., 1997).

Fig. 3.2.9. *Flow past a circular cylinder. Lattice with 6400 times 2560 nodes. The uppermost plot shows the result after 20000 time steps for the FHP-I model (the flow velocity averaged over the whole domain has been substracted in both plots in order to make the pertubations better visible). If 3-particle collisions are left out the flow field looks quite different (lower plot).*

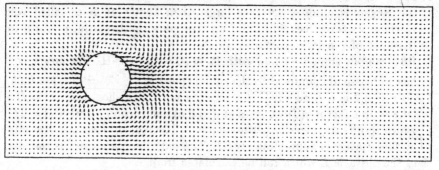

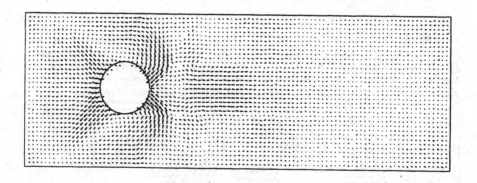

Exercise 3.2.10. (*)
Consider typical flow velocities in the atmosphere and oceans. At which spatial scales the Reynolds numbers are 1, 1000 and 10^6?

3.2.10 The 8-bit FHP model

D'Humières et al. (1987) proposed a model with two populations (n_0, m_0) that represent rest particles. With 2 bits up to 3 rest particles can be described: (n_0, m_0) = (0,0) no, (1,0) one, (0,1) two and (1,1) three rest particles. All collisions conserving mass and momentum are included ('collision saturated'). The collisions a) to c) in Table 3.2.2 proceed independent of the number of rest particles. The collisions leading to creation and destruction of rest particles are all included, except a few cases which take place with probability x, y, or z (compare Tables 3.2.10 and 3.2.11). It can be shown that for $x = 0.5$, $y = z = 0.2$, there exists a value for d for which $g(d) = 1$ and $dg(d)/dd = 0$. D'Humières et al. recommend a model with $x = 1/2$, $y = z = 1/6$ which gives $g = 1.0$ at $d = 0.21$. The model violates semi-detailed balance (compare Exercise 3.2.11). Numerical experiments show that vorticity is advected close to the flow speed U_0. The 8-bit-FHP model is probably already too complex to allow coding with bit-operators. The use of look-up tables makes it much more clumpsy than the 7-bit-FHP models.

Exercise 3.2.11. (*)
Show that the 8-bit FHP model of d'Humières et al. (1987) violates semi-detailed balance.

Exercise 3.2.12. (***)
Repeat the simulation of the flow past a cylinder with the FHP-II model. Compare the computational time for constant Reynolds number by varying the upstream velocity or the size of the system. Analyze the differences in the flow pattern when applying the random or the chiral two-particle head-on collision.

Fig. 3.2.10. *FHP-8-bit model: collisions involving rest particles; occupied cells are represented by arrows, empty cells by thin lines, the number of rest particles is indicated by the number in the central circle.*

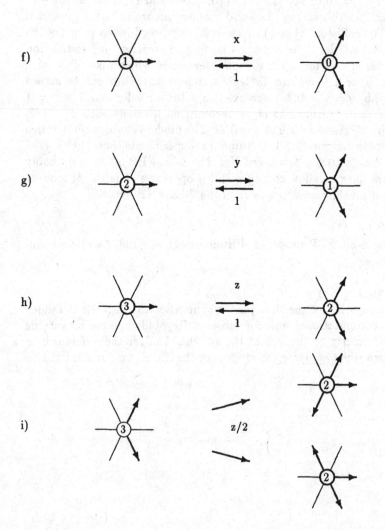

Fig. 3.2.11. ***FHP-8-bit model: collisions involving rest particles (continued); occupied cells are represented by arrows, empty cells by thin lines, the number of rest particles is indicated by the number in the central circle.***

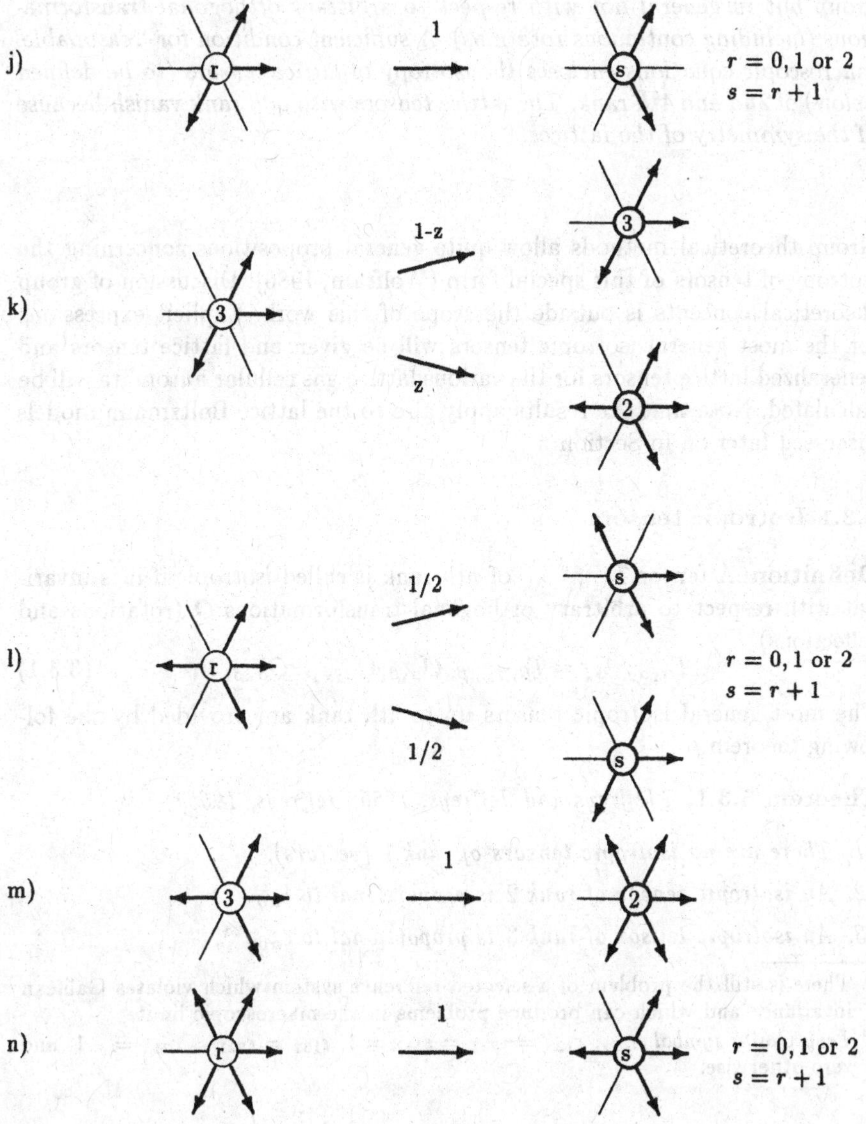
j)
r
1
s
r = 0, 1 or 2
s = r + 1
k)
3
1-z
z
3
2
l)
r
1/2
1/2
s
s
r = 0, 1 or 2
s = r + 1
m)
3
1
2
n)
r
1
s
r = 0, 1 or 2
s = r + 1

3.3 Lattice tensors and isotropy in the macroscopic limit

All lattice-gas cellular automata are based on extremely discretized phase spaces. Nevertheless it is to be expected that the spatial discretization will be smoothed out on scales which are much larger than the grid spacing[27]. Things are less obvious for the discretization of the velocities. Calculation of mean values over an angular direction is restricted to a range of 2π which contains only a few velocities (six for FHP). The multi-spin coding yields tensors which consist of the components of the lattice velocities. These tensors are invariant with respect to elements of the associated finite symmetry group but in general not with respect to arbitrary orthogonal transformations (including continuous rotations). A sufficient condition for 'reasonable' macroscopic equations encloses the isotropy of lattice tensors (to be defined below) of 2nd and 4th rank. The lattice tensors with odd rank vanish because of the symmetry of the lattices.

Group theoretical methods allow quite general propositions concerning the isotropy of tensors of this special form (Wolfram, 1986). Discussion of group theoretical concepts is outside the scope of this work. Explicit expressions for the most general isotropic tensors will be given and lattice tensors and generalized lattice tensors for the various lattice-gas cellular automata will be calculated. Note that the results apply also to the lattice Boltzmann models discussed later on in Section 5.

3.3.1 Isotropic tensors

Definition: A tensor $T_{\alpha_1\alpha_2...\alpha_n}$ of nth rank is called *isotropic* if it is invariant with respect to arbitrary orthogonal transformations O (rotations and reflections)

$$T_{\alpha_1\alpha_2...\alpha_n} = T_{\beta_1\beta_2...\beta_n}O_{\alpha_1\beta_1}O_{\alpha_2\beta_2}...O_{\alpha_n\beta_n}. \qquad (3.3.1)$$

The most general isotropic tensors up to 4th rank are provided by the following theorem.

Theorem 3.3.1. *(Jeffreys and Jeffreys, 1956; Jeffreys, 1965)*

1. *There are no isotropic tensors of rank 1 (vectors).*

2. *An isotropic tensor of rank 2 is proportional to $\delta_{\alpha\beta}$.*

3. *An isotropic tensor of rank 3 is proportional to $\epsilon_{\alpha\beta\gamma}$[28].*

---

[27] There is still the problem of a selected reference system which violates Galilean invariance and which can produce problems in the macroscopic limit.

[28] *Levy-Civita symbol $\epsilon_{\alpha\beta\gamma}$:* $\epsilon_{123} = \epsilon_{231} = \epsilon_{312} = 1$, $\epsilon_{132} = \epsilon_{321} = \epsilon_{213} = -1$, and zero otherwise.

4. There are three different (linear independent) tensors of rank 4

$$\delta_{\alpha\beta}\delta_{\gamma\delta}, \quad \delta_{\alpha\gamma}\delta_{\beta\delta}, \quad \delta_{\alpha\delta}\delta_{\beta\gamma},$$

which can be combined to the most general form

$$T_{\alpha\beta\gamma\delta} = a\delta_{\alpha\beta}\delta_{\gamma\delta} + b\delta_{\alpha\gamma}\delta_{\beta\delta} + c\delta_{\alpha\delta}\delta_{\beta\gamma}, \qquad (3.3.2)$$

where a, b and c are arbitrary constants.

A proof of the theorem can be found in Jeffreys (1965).

Isotropic tensors of rank $n \geq 4$ consist only of products of second rank δ tensors (for example: $\delta_{\alpha\beta}\delta_{\gamma\delta}\delta_{\epsilon\zeta}$ and all tensors that result from cyclic permutation of indices) when n is even or of products of δ and ϵ tensors when n is odd.

In two dimensions the isotropic tensor of rank 4 (3.3.2) has the following non-vanishing components:

$$
\begin{aligned}
T_{1111} = T_{2222} &= a + b + c, \\
T_{1122} = T_{2211} &= a, \\
\\
T_{1212} = T_{2121} &= b, \\
T_{1221} = T_{2112} &= c.
\end{aligned}
\qquad (3.3.3)
$$

In particular, the tensor $\delta_{\alpha\beta\gamma\delta}$ is non-isotropic ($\delta_{\alpha\beta\gamma\delta}$ is 1 if all indices are equal and 0 otherwise; it is a generalization of the Kronecker symbol $\delta_{\alpha\beta}$). The same is true for $\delta_{\alpha\beta\gamma\delta\epsilon\zeta}$.

3.3.2 Lattice tensors: single-speed models

Let us define the following tensors of rank n

$$\boxed{L_{\alpha_1\alpha_2...\alpha_n} = \sum_i c_{i\alpha_1} c_{i\alpha_2} ... c_{i\alpha_n}} \qquad (3.3.4)$$

where $c_{i\alpha_\nu}$ are the cartesian components of the lattice velocities c_i. We will call these tensors the *lattice tensors*. Because of their special structure they are invariant with respect to the symmetry group of the lattice and they are symmetric in all of their indices. From these symmetries it follows that these tensors can have a maximal number N of independent components of

$$N = \binom{n + D - 1}{n} = \frac{(n + D - 1)!}{n!(D - 1)!}$$

where D is the (spatial) dimension. Example: A symmetric tensor T of rank 2 in two dimensions has at most $N = \binom{3}{2} = \frac{3!}{2!1!} = 3$ independent components:

$$T = \begin{pmatrix} a & b \\ b & c \end{pmatrix}.$$

In lattice-gas cellular automata with one speed (HPP, FHP, FCHC) the momentum advection tensor (MAT) of 4th rank $T_{\alpha\beta\gamma\delta}^{(MA)}$ (compare Eq. 3.2.55) occurs in the macroscopic form of the momentum balance. It can be rewritten in terms of the lattice tensors of rank two and four:

$$
\begin{aligned}
T_{\alpha\beta\gamma\delta}^{(MA)} &= \sum_i c_{i\alpha} c_{i\beta} Q_{i\gamma\delta} \\
&= \sum_i \left[c_{i\alpha} c_{i\beta} c_{i\gamma} c_{i\delta} - \frac{1}{2} c_{i\alpha} c_{i\beta} \delta_{\gamma\delta} \right] \\
&= L_{\alpha\beta\gamma\delta} - \frac{1}{2} L_{\alpha\beta} \delta_{\gamma\delta}.
\end{aligned}
\tag{3.3.5}
$$

A sufficient condition for the isotropy of $T_{\alpha\beta\gamma\delta}^{(MA)}$ is the isotropy of $L_{\alpha\beta}$ and $L_{\alpha\beta\gamma\delta}$. $T_{\alpha\beta\gamma\delta}^{(MA)}$ is non-isotropic if $L_{\alpha\beta}$ is isotropic while $L_{\alpha\beta\gamma\delta}$ is non-isotropic. Other combinations do not occur in the models considered. In what follows we use the notation $DkQb$ of Qian et al. (1992) where k is the spatial dimension and b is the number of lattice velocities (including $c_0 = 0$ for rest particles).

Square lattice: HPP (D2Q4). Lattice velocities:

$$c_i = \left(\cos \frac{2\pi i}{4}, \sin \frac{2\pi i}{4} \right) \quad i = 1, 2, 3, 4.$$

The lattice tensor of rank 2

$$L_{\alpha\beta}^{HPP} = 2 \begin{pmatrix} 1 & 0 \\ 0 & 1 \end{pmatrix} = 2\delta_{\alpha\beta}$$

is isotropic.
The lattice tensor of rank 4

$$L_{\alpha\beta\gamma\delta}^{HPP} = 2\,\delta_{\alpha\beta\gamma\delta} \tag{3.3.6}$$

is non-isotropic. As a consequence the HPP model fails to yield the Navier-Stokes equations in the macroscopic limit.

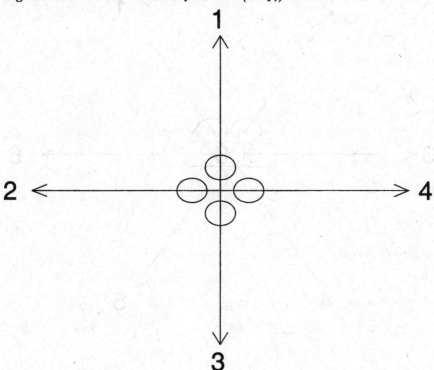

Fig. 3.3.1. *The lattice velocities of the HPP (D2Q4) lattice.*

Triangular lattice: FHP (D2Q7). The triangular FHP lattice has hexagonal symmetry. Lattice velocities:

$$c_0 = (0,0)$$
$$c_i = \left(\cos \frac{2\pi i}{6}, \sin \frac{2\pi i}{6} \right) \quad i = 1, ..., 6 \qquad (3.3.7)$$

Fig. 3.3.2. *The lattice velocities of the FHP (D2Q7) lattice.*

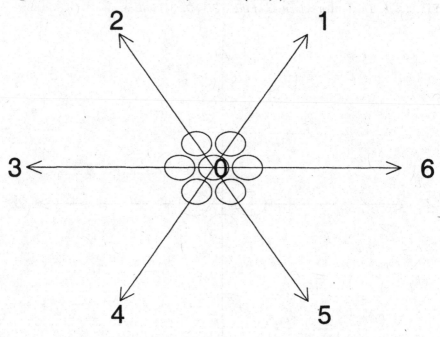

The lattice tensors of rank 2

$$L_{\alpha\beta}^{FHP} = 3\,\delta_{\alpha\beta} \tag{3.3.8}$$

and rank 4

$$L_{\alpha\beta\gamma\delta}^{FHP} = \frac{3}{4}\left(\delta_{\alpha\beta}\delta_{\gamma\delta} + \delta_{\alpha\gamma}\delta_{\beta\delta} + \delta_{\alpha\delta}\delta_{\beta\gamma}\right) \tag{3.3.9}$$

are isotropic. The momentum advection tensor reads

$$T_{\alpha\beta\gamma\delta}^{(MA)} = \frac{3}{4}\left(\delta_{\alpha\gamma}\delta_{\beta\delta} + \delta_{\alpha\delta}\delta_{\beta\gamma} - \delta_{\alpha\beta}\delta_{\gamma\delta}\right). \tag{3.3.10}$$

Thus, FHP yields the Navier-Stokes equation in the macroscopic limit.

FCHC (D4Q24). The investigation of the lattice tensors of the FCHC model is left to the reader (Exercise 3.3.1).

Exercise 3.3.1. (**)
The 24 lattice velocities of the FCHC model are given by

$$(\pm 1, \pm 1, 0, 0), \quad (\pm 1, 0, \pm 1, 0), \quad (0, \pm 1, \pm 1, 0),$$
$$(\pm 1, 0, 0, \pm 1), \quad (0, \pm 1, 0, \pm 1), \quad (0, 0, \pm 1, \pm 1).$$

Calculate the lattice tensors of rank 2 and 4 and show that they are isotropic.

3.3.3 Generalized lattice tensors for multi-speed models

Later on multi-speed[29] lattice-gas cellular automata and multi-speed lattice Boltzmann models will be discussed. The associated lattice tensors of rank 4 are usually non-isotropic because the symmetry group of the corresponding lattices is not large enough (note that the situation is different for the multi-speed FHP model where the symmetry of each single-speed sub-lattice is large enough). But isotropy of 4th rank tensors can be recovered by introducing weights w_i for the different speeds. These *generalized lattice tensors*

$$\boxed{G_{\alpha_1\alpha_2...\alpha_n} = \sum_i w_i c_{i\alpha_1} c_{i\alpha_2}...c_{i\alpha_n}} \tag{3.3.11}$$

occur naturally in the multi-scale analysis of multi-speed models. The weights correspond to different occupation numbers for the different speeds in the global equilibrium with vanishing macroscopic velocities.
The following sub-sections can be skipped in the first reading and should be revisited when the appropriate multi-speed model is discussed.

---

[29] Models with several different speeds have been encountered before, namely the FHP variants with rest particles (FHP-II, FHP-III). Particles with vanishing speed have no influence on the isotropy of the lattice tensors. Therefore only models with different *non-vanishing* speeds are called multi-speed models.

D2Q9. Lattice velocities:

$$
\begin{aligned}
c_0 &= (0,0), \\
c_{1,3}, c_{2,4} &= (\pm 1, 0), \quad (0, \pm 1), \\
c_{5,6,7,8} &= (\pm 1, \pm 1).
\end{aligned}
\tag{3.3.12}
$$

Fig. 3.3.3. *The lattice velocities of the D2Q9 lattice.*

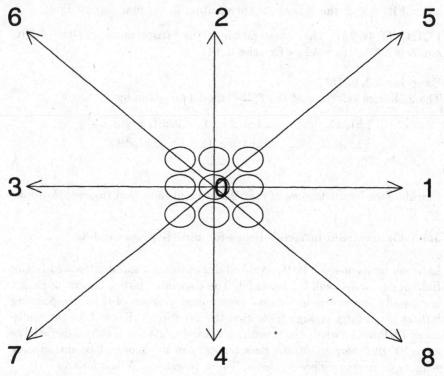

The lattice tensor of rank 2

$$
L_{\alpha\beta}^{(D2Q9)} = 6\delta_{\alpha\beta}
$$

is isotropic. The lattice tensor of rank 4 with the following non-vanishing components

$$
\begin{aligned}
L_{1111}^{(D2Q9)} = L_{2222}^{(D2Q9)} &= 6 \\
L_{1122}^{(D2Q9)} = L_{2211}^{(D2Q9)} &= 4 \\
L_{1212}^{(D2Q9)} = L_{2121}^{(D2Q9)} &= 4 \\
L_{1221}^{(D2Q9)} = L_{2112}^{(D2Q9)} &= 4
\end{aligned}
$$

is non-isotropic.

Introducing the weights $w_i = 1$ for speed 1 ($i = 1, ..., 4$) and $w_i = 1/4$ for speed $\sqrt{2}$ ($i = 5, ..., 8$) leads to isotropic generalized lattice tensors of rank 2

$$G_{\alpha\beta}^{(D2Q9)} = 3\delta_{\alpha\beta} \tag{3.3.13}$$

and rank 4

$$G_{\alpha\beta\gamma\delta}^{(D2Q9)} = \delta_{\alpha\beta}\delta_{\gamma\delta} + \delta_{\alpha\gamma}\delta_{\beta\delta} + \delta_{\alpha\delta}\delta_{\beta\gamma}. \tag{3.3.14}$$

D2Q13-WB. Weimar and Boon (1996)

$$
\begin{aligned}
c_0 &= (0,0) & \text{rest particle} \\
c_{1,2},\, c_{3,4} &= (\pm 1, 0),\, (0, \pm 1) & \text{1-particles} \\
c_{5,6,7,8} &= (\pm 1, \pm 1) & \sqrt{2}\text{-particles} \\
c_{9,10},\, c_{11,12} &= (\pm 2, 0),\, (0, \pm 2) & \text{2-particles}
\end{aligned}
\tag{3.3.15}
$$

Fig. 3.3.4. *The lattice velocities of the D2Q13-WB lattice.*

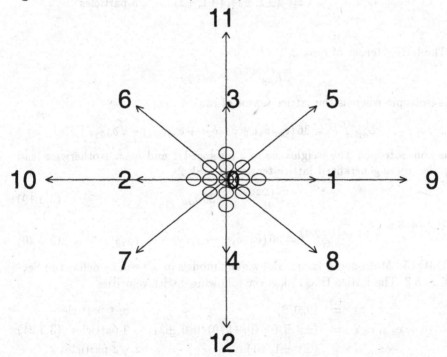

The lattice tensor of rank 2

$$L_{\alpha\beta}^{(D2Q13-WB)} = 14\,\delta_{\alpha\beta}$$

is isotropic whereas the lattice tensor of rank 4

$$L_{\alpha\beta\gamma\delta}^{(D2Q13-WB)} = 4(\delta_{\alpha\beta}\delta_{\gamma\delta} + \delta_{\alpha\gamma}\delta_{\beta\delta} + \delta_{\alpha\delta}\delta_{\beta\gamma}) + 26\,\delta_{\alpha\beta\gamma\delta}$$

is non-isotropic. The weights $w_i = 4$ for speed 1, $w_i = 5$ for speed $\sqrt{2}$, and $w_i = 1$ for speed 2 lead to isotropic generalized lattice tensors of rank 2

$$G_{\alpha\beta}^{(D2Q13-WB)} = 36\,\delta_{\alpha\beta} \tag{3.3.16}$$

and rank 4

$$G_{\alpha\beta\gamma\delta}^{(D2Q13-WB)} = 20(\delta_{\alpha\beta}\delta_{\gamma\delta} + \delta_{\alpha\gamma}\delta_{\beta\delta} + \delta_{\alpha\delta}\delta_{\beta\gamma}). \tag{3.3.17}$$

D2Q21. The D2Q21 lattice was introduced by Fahner (1991).

$$
\begin{aligned}
c_0 &= (0,0) & &\text{rest particle} \\
c_{1,2},\ c_{3,4} &= (\pm 1, 0),\ (0, \pm 1) & &\text{1-particles} \\
c_{5,6,7,8} &= (\pm 1, \pm 1) & &\sqrt{2}\text{-particles} \\
c_{9,10},\ c_{11,12} &= (\pm 2, 0),\ (0, \pm 2) & &\text{2-particles} \\
c_{13,\dots,16},\ c_{17,\dots,20} &= (\pm 2, \pm 1),\ (\pm 1, \pm 2) & &\sqrt{5}\text{-particles}
\end{aligned} \tag{3.3.18}
$$

The lattice tensor of rank 2

$$L_{\alpha\beta}^{(D2Q21)} = 34\,\delta_{\alpha\beta}$$

is isotropic whereas the lattice tensor of rank 4

$$L_{\alpha\beta\gamma\delta}^{(D2Q21)} = 36\,(\delta_{\alpha\beta}\delta_{\gamma\delta} + \delta_{\alpha\gamma}\delta_{\beta\delta} + \delta_{\alpha\delta}\delta_{\beta\gamma}) - 2\,\delta_{\alpha\beta\gamma\delta}$$

is non-isotropic. The weights $w_i = 2$ for speed 1 and $w_i = 1$ otherwise lead to isotropic generalized lattice tensors of rank 2

$$G_{\alpha\beta}^{(D2Q21)} = 36\,\delta_{\alpha\beta} \tag{3.3.19}$$

and rank 4

$$G_{\alpha\beta\gamma\delta}^{(D2Q21)} = 36\,(\delta_{\alpha\beta}\delta_{\gamma\delta} + \delta_{\alpha\gamma}\delta_{\beta\delta} + \delta_{\alpha\delta}\delta_{\beta\gamma}) \tag{3.3.20}$$

D3Q15. Multi-speed lattice Boltzmann models in 3D will be defined in Section 5.3. The lattice D3Q15 has the following lattice velocities

$$
\begin{aligned}
c_0 &= (0,0,0) & &\text{rest particle} \\
c_{1,2},\ c_{3,4},\ c_{5,6} &= (\pm 2, 0, 0),\ (0, \pm 2, 0)\ (0, 0, \pm 2) & &\text{2-particles} \\
c_{7,\dots,14} &= (\pm 1, \pm 1, \pm 1) & &\sqrt{3}\text{-particles}
\end{aligned} \tag{3.3.21}
$$

Fig. 3.3.5. *The lattice velocities of the D2Q21 lattice.*

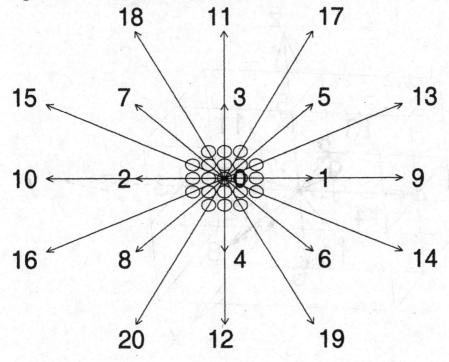

Fig. 3.3.6. *The lattice velocities of the D3Q15 lattice.*

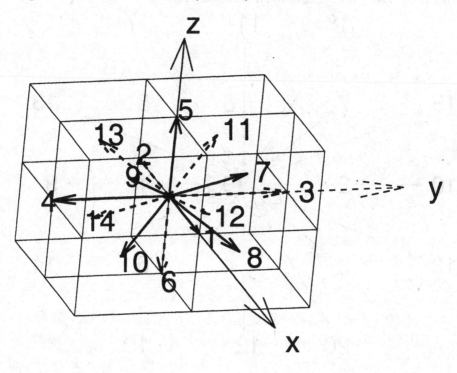

The lattice tensor of rank 2

$$L_{\alpha\beta}^{(D3Q15)} = 16\,\delta_{\alpha\beta}$$

is isotropic whereas the lattice tensor of rank 4

$$L_{\alpha\beta\gamma\delta}^{(D3Q15)} = 8(\delta_{\alpha\beta}\delta_{\gamma\delta} + \delta_{\alpha\gamma}\delta_{\beta\delta} + \delta_{\alpha\delta}\delta_{\beta\gamma}) + 16\,\delta_{\alpha\beta\gamma\delta}$$

is non-isotropic. The weights $w_i = 2$ for speed $\sqrt{3}$ and $w_i = 1$ for speed 2 lead to isotropic generalized lattice tensors of rank 2

$$G_{\alpha\beta}^{(D3Q15)} = 24\,\delta_{\alpha\beta} \tag{3.3.22}$$

and rank 4

$$G_{\alpha\beta\gamma\delta}^{(D3Q15)} = 16\,(\delta_{\alpha\beta}\delta_{\gamma\delta} + \delta_{\alpha\gamma}\delta_{\beta\delta} + \delta_{\alpha\delta}\delta_{\beta\gamma}). \tag{3.3.23}$$

D3Q19. The model D3Q19 has the following lattice velocities

$$
\begin{aligned}
c_0 &= (0,0) \\
c_{1,2},\ c_{3,4},\ c_{5,6} &= (\pm1,0,0),\ (0,\pm1,0)\ (0,0,\pm1) \\
c_{7,\ldots,10},\ c_{11,\ldots,14},\ c_{15,\ldots,18} &= (\pm1,\pm1,0),\ (\pm1,0,\pm1),\ (0,\pm1,\pm1).
\end{aligned}
\tag{3.3.24}
$$

The lattice tensor of rank 2

$$L_{\alpha\beta}^{(D3Q19)} = 10\,\delta_{\alpha\beta}$$

is isotropic whereas the lattice tensor of rank 4

$$L_{\alpha\beta\gamma\delta}^{(D3Q19)} = 4\,(\delta_{\alpha\beta}\delta_{\gamma\delta} + \delta_{\alpha\gamma}\delta_{\beta\delta} + \delta_{\alpha\delta}\delta_{\beta\gamma}) - 2\,\delta_{\alpha\beta\gamma\delta}$$

is non-isotropic. The weights $w_i = 2$ for speed 1 and $w_i = 1$ for speed $\sqrt{2}$ lead to isotropic generalized lattice tensors of rank 2

$$G_{\alpha\beta}^{(D3Q19)} = 12\,\delta_{\alpha\beta} \tag{3.3.25}$$

and rank 4

$$G_{\alpha\beta\gamma\delta}^{(D3Q19)} = 4\,(\delta_{\alpha\beta}\delta_{\gamma\delta} + \delta_{\alpha\gamma}\delta_{\beta\delta} + \delta_{\alpha\delta}\delta_{\beta\gamma}). \tag{3.3.26}$$

3.3.4 Thermal LBMs: D2Q13-FHP (multi-speed FHP model)

For thermal lattice Boltzmann models (Navier-Stokes plus energy equation) isotropic lattice tensors up to rank 6 are required.

$$
\begin{aligned}
c_i &= (0,0) & i &= 0 \\
c_i &= \left(\cos\frac{2\pi k}{6},\ \sin\frac{2\pi k}{6}\right) & i &= 1,2,\ldots,6;\quad k = i \\
c_i &= 2\left(\cos\frac{2\pi k}{6},\ \sin\frac{2\pi k}{6}\right) & i &= 7,8,\ldots,12;\quad k = i - 6
\end{aligned}
\tag{3.3.27}
$$

Fig. 3.3.7. *The lattice velocities of the D3Q19 lattice.*

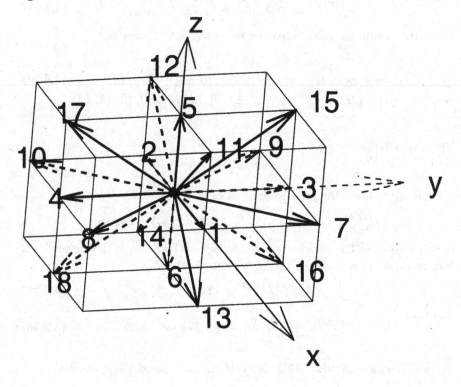

Fig. 3.3.8. *The lattice velocities of the D2Q13-FHP (multi-speed FHP) lattice.*

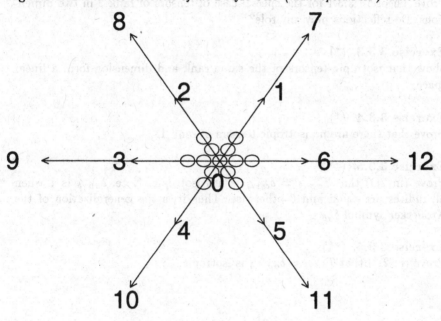

The lattice tensors of rank 2

$$L_{\alpha\beta}^{(D2Q13-FHP)} = 51\delta_{\alpha\beta} \tag{3.3.28}$$

and of rank 4

$$L_{\alpha\beta\gamma\delta}^{(D2Q13-FHP)} = 15(\delta_{\alpha\beta}\delta_{\gamma\delta} + \delta_{\alpha\gamma}\delta_{\beta\delta} + \delta_{\alpha\delta}\delta_{\beta\gamma}) \tag{3.3.29}$$

are isotropic.

3.3.5 Exercises

Exercise 3.3.2. (*)
Prove Theorem 3.3.1 for the special case of tensors of rank 2 in two dimensions. Do reflections play any role?

Exercise 3.3.3. (*)
Show that isotropic tensors of the same rank and dimension form a linear space.

Exercise 3.3.4. (*)
Prove that there are no isotropic tensors of rank 1.

Exercise 3.3.5. (**)
Prove (in 2D) that $T_{\alpha\beta\gamma\delta} = \delta_{\alpha\beta\gamma\delta}$ is not isotropic. Note: $\delta_{\alpha\beta\gamma\delta}$ is 1 when all indices are equal and 0 otherwise. Thus it is the generalization of the Kronecker symbol $\delta_{\alpha\beta}$.

Exercise 3.3.6. (**)
Prove (in 2D) that $T_{\alpha\beta\gamma\delta} = \delta_{\alpha\beta}\delta_{\gamma\delta}$ is isotropic.

3.4 Desperately seeking a lattice for simulations in three dimensions

Frisch, Hasslacher and Pomeau (1986) found out that in addition to mass and momentum conservation lattice-gas cellular automata for the Navier-Stokes equation must reside on a grid with 'sufficient' symmetry. The importance of this insight can hardly be overestimated. Thus the first task in the development of a LGCA for simulations in 3D is to find a lattice with appropriate symmetry. This is not as easy as in 2D.

3.4.1 Three dimensions

In close analogue to 2D where lattice vectors c_i were defined by the corners of regular polygons let us define c_i in 3D by the corners of regular polytopes. In three dimensions only a few regular polytopes, namely the *Platonic solids*[30], lead to lattice vectors of equal length. There are exactly five Platonic solids: tetrahedron, hexahedron (cube), octahedron, dodecahedron and icosahedron (compare Table 3.4.1 and Fig. 3.4.1).

The cube and the octahedron are *dual* to each other in the following sense: the mid-points of the faces of a cube yield the corners of an octahedron and the mid-points of an octahedron yield the corners of a cube. In the same sense dodecahedron and icosahedron are dual to each other. The tetrahedron is self-dual: its dual solid is also a tetrahedron.

Rotations and reflections which transform the solid onto itself form a group which is referred to as the symmetry group of the solid. Each rotation or reflection which transforms the solid onto itself does the same thing with the dual solid embedded. Therefore dual regular polytopes show identical symmetries and their corresponding symmetry groups are isomorphic (Ledermann, 1985a). There may be further polytopes with the same symmetry group which, however, are not regular.

In order to be useful as building blocks of a lattice-gas cellular automata, a polyhedron must show a large enough symmetry group (this constraint is fulfilled only by the dodecahedron and its dual partner the icosahedron) and in addition must fill the whole space[31]. The cube is the only Platonic solid whichs completely fills the space without gaps (Ledermann, 1985a). Thus in 3D there is no polyhedron which respects all constraints.

---

[30] Definition of Platonic solids: convex polytopes, bounded by regular congruent polyhedrons and with equal number of edges meeting at each corner.

[31] Actually it is required that all corners (which are nodes of the lattices) are connected to an equal number of nodes by the lattice vectors c_i. Gaps between the polytopes could lead to nodes with a smaller number of nearest neighbors. As an analog in 2D consider the parqueting of the plane with octagons.

Fig. 3.4.1. *The five Platonic solids.*

Tetrahedron

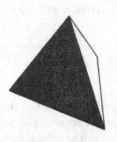

Hexahedron (Cube)

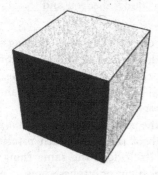

Octahedron

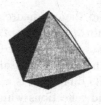

Dodecahedron

Icosahedron

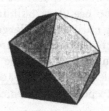

Table 3.4.1. *The five Platonic solids: number of faces F, edges E and corners C.*
Euler's polyhedron theorem: F − E + C = 2

polyhedron	F	E	C	dual solid
tetrahedron	4	6	4	tetrahedron
hexahedron (cube)	6	12	8	octrahedron
octrahedron	8	12	6	hexahedron (cube)
dodecahedron	12	30	20	icosahedron
icosahedron	20	30	12	dodecahedron

Table 3.4.2. *Lattice tensors of ranks 2 to 6 for the five Platonic solids: isotropic*
(+) or not (−) (from Wolfram, 1986). The number of lattice vectors is equal to the
number of corners C.

polyhedron	C	2	3	4	5	6
tetrahedron	4	+	−	−	−	−
hexahedron (cube)	8	+	+	−	+	−
octrahedron	6	+	+	−	+	−
dodecahedron	20	+	+	+	+	−
icosahedron	12	+	+	+	+	−

Historical remark: In 'Mysterium Cosmographicum' (1596) Johannes Kepler (1571-1630) has suggested that the distances of the then known planets - Mercury to Saturn - are constrained by the Platonic solids which are alternatingly inscribed and circumscribed to spheres. (compare Fig. 3.4.2).

Fig. 3.4.2. *In 'Mysterium Cosmographicum' (1596) Johannes Kepler (1571-1630) has suggested that the distances of the then known planets - Mercury to Saturn - are constrained by the Platonic solids which are alternatingly inscribed and circumscribed to spheres. The six spheres correspond to the six planets, Saturn, Jupiter, Mars, Earth, Venus, Mercurius, separated in the order by cube, tetrahedron, dodecahedron, octahedron and icosahedron (adapted from Weyl, 1989).*

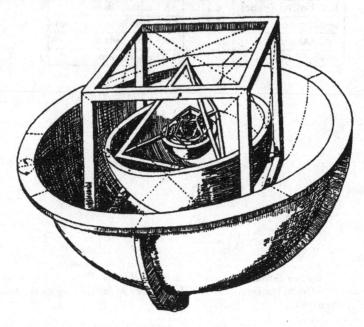

3.4.2 Five and higher dimensions

A further possibility is to find lattices in higher dimensions with sufficient symmetry. The quantities of interest can then be obtained by appropriate projections from higher dimensions down to 3D. In five and higher dimensions there are only three regular polytopes[32] for each dimension, namely

---

[32] These three regular polytopes exist also in lower dimensions.

the *simplex*[33], the *hypercube*[34] and its dual solid. The corresponding lattice tensors of rank 4 are isotropic only for $D < 3$ for the simplex and only for $D < 4$ for the hypercube and its dual solid.

3.4.3 Four dimensions

The last chance lies in four dimensions where in addition to the simplex, the hypercube and its dual solid there exist further three regular polytopes which can be characterized by the Schläfli symbols[35] $\{3, 4, 3\}$, $\{3, 3, 5\}$, and $\{5, 3, 3\}$. The $\{3, 4, 3\}$-polytop is referred to as *face-centered hypercube* (FCHC). It has 24 corners with coordinates which are permutations of $(\pm 1, \pm 1, 0, 0)$. The corresponding lattice tensors are isotropic up to 4th rank inclusively. Thus FCHC is the lattice searched for! The 24 lattice vectors of FCHC read

$$(\pm 1, \pm 1, 0, 0), \quad (\pm 1, 0, \pm 1, 0), \quad (\pm 1, 0, 0, \pm 1),$$

$$(3.4.1)$$

$$(0, \pm 1, \pm 1, 0), \quad (0, \pm 1, 0, \pm 1), \quad (0, 0, \pm 1, \pm 1).$$

The projections of FCHC into 3D space are shown in Figures 3.4.3 and 3.4.4. There are 12 velocities with $c_{i4} = 0$ (Fig. 3.4.3) and two times 6 velocities with $c_{i4} = \pm 1$ (Fig. 3.4.4).

The other regular polytopes $\{3, 3, 5\}$ and $\{5, 3, 3\}$ are dual to each other. The $\{3, 3, 5\}$ polytope has 120 corners and the corresponding lattice tensors are isotropic up to 8th rank inclusively. The $\{3, 3, 5\}$ and $\{5, 3, 3\}$ polytopes have not been mentioned by Wolfram (1986) as alternatives for applications in lattice-gas automata. One reason at least is simplicity: FCHC has much less corners than the $\{3, 3, 5\}$ polytop.

Further reading on regular polytopes: Coxeter (1963).

---

[33] The set of all vectors x which respect the constraints

$$x = \lambda_1 x_1 + \ldots + \lambda_r x_r, \quad x_i \in \mathbf{R}^D, \quad \lambda_i \in \mathbf{R}, \quad \lambda_i \geq 0, \quad \sum_{i=1}^{r} \lambda_i = 1,$$

are referred to as simplex $[x_1, ..., x_r]$. Despite of some degenerated cases 2, 3, and 4 points lead to a line segment, a (planar) triangle and a tetrahedron as simplices (Ledermann, 1985a, p. 108).

[34] The hypercube γ_D in D dimensions has 2^D corners $\pm e_1 \pm \ldots \pm e_D$. Its $2 \cdot D$ 'surfaces' are $(D-1)$-dimensional hypercubes. γ_1 is a line segment, γ_2 a square, and γ_3 a cube (Ledermann, 1985b). $e_i = (0, ..., 0, 1, 0, ..., 0)$ are the standard base vectors.

[35] Compare Appendix 6.5.

Fig. 3.4.3. *The projection (along the 4th axis) of the lattice velocities of the FCHC lattice for $c_{i4} = 0$.*

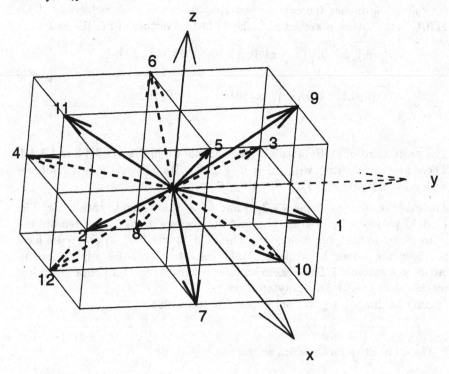

Fig. 3.4.4. *The projection (along the 4th axis) of the lattice velocities of the FCHC lattice for $c_{i4} = \pm 1$.*

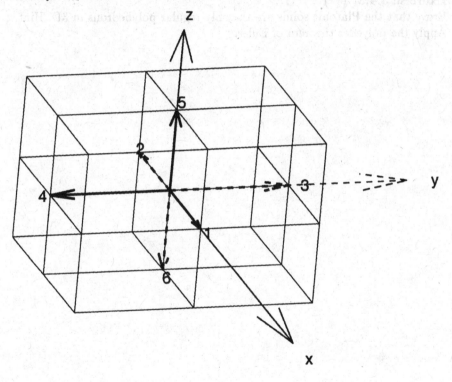

Exercise 3.4.1. (***)
Is it possible to build a space-filling lattice from the regular polytop $\{3,3,5\}$ alone?

Exercise 3.4.2. (***)
Prove the polyeder theorem of Leonard Euler (1707-1783): The number C of corners plus the number F of faces equals the number E of edges plus 2:

$$C + F = E + 2.$$

Exercise 3.4.3. (**)
Prove that the Platonic solids are the only regular polyhedrons in 3D. Hint: Apply the polyeder theorem of Euler.

3.5 FCHC

D'Humières, Lallemand and Frisch (1986) proposed the face-centered hyper-cube (FCHC) as a lattice with sufficient symmetry for hydrodynamic sim-ulations in 3D but without specifying collision rules. Those were given for the first time by Hénon (1987a) and later on modified by Rem and Somers (1989), Somers and Rem (1989) and others.

The 24 vectors of the FCHC lattice are listed in Section 3.3. The collision rules have to respect the following constraints:

1. The number of particles is conserved in each collision (conservation of mass).

2. The momentum is conserved in each collision.

3. There are no conserved quantities except of mass and momentum (no spurious invariants).

4. The exclusion principle is valid: at each node there sits at most one particle per lattice velocity (or per cell).

5. The collision rules share the symmetry of the lattice or, in other words, they are invariant under arbitrary transformations by elements of the isometric (symmetry) group $\mathcal{G}$.

6. The collisions respect the semi-detailed balance.

The FHP-model with six lattice velocities (FHP-I) has only $2^6 = 64$ different states per node. Therefore the collision rules could be derived 'by hand'. On the contrary, for FCHC there are $2^{24} = 16\,777\,216$ different states. Thus the collision rules have to be specified by an automatic algorithm.

3.5.1 Isometric collision rules for FCHC by Hénon

Despite the constraints given above there is space for almost unnumerable many different collisions. Therefore Hénon (1989a) introduced further con-straints (referred to as *Hénon constraints*):

1. Every collision is an isometry[36] (isometric collision rules). Motivation:

   – An isometry is simpler than an arbitrary transformation.

   – The general constraints 1 and 4 are fulfilled automatically.

---

[36] Isometries are mappings g which keep the distances $d(\alpha, \beta)$ of arbitrary points α, β invariant: $d(g(\alpha), g(\beta)) = d(\alpha, \beta)$. Rotations and reflections are isometries.

- All collisions of HPP and FHP (without rest particles) are isometries.

2. The isometry depends only on the total momentum.

3. The isometry which is to be applied will be chosen by random out of the optimal (with respect to low viscosity, see below) isometries.

Despite the first two Hénon constraints there exists in addition to the identity at least one nontrivial isometry for each total momentum.

The various isometries contribute differently to the shear viscosity which should be kept as low as possible. Hénon could quantify these contributions and thereby classify the optimal isometries. This isometric group G contains 1152 elements. It can be created by five generators.

In order to keep the collisions easy to take at a glance, Hénon introduced a kind of normal form of the components of momentum q_1, q_2, q_3, q_4. Note that this is not a further constraint to the collision rules. For every state there exists an isometry which transforms that state into a state (*normalized momenta*) with

$$q_1 \geq q_2 \geq q_3 \geq q_4 \quad \text{and} \quad (q_4 = 0 \quad \text{or} \quad q_1 + q_4 < q_2 + q_3). \tag{3.5.1}$$

Thereby the problem of constructing collision rules for FCHC is simplified considerably because there exist only 37 different normalized momenta.

Each collision proceeds by successive application of three isometries:

1. Transformation of the initial state by an isometry $g \in G$ into the state with the appropriate normalized momentum.

2. Application of an optimal isometry M (proper collision).

3. Transformation with $g^{-1} \in G$ back to the original coordinates.

Thus the total transformation reads $g^{-1} \circ M \circ g$.

3.5.2 FCHC, computers and modified collision rules

Despite the extreme simplification of the original problem of constructing collision rules for FCHC by Hénon the coding of these rules require the introduction of a very large *look-up table* that contains the final state (after collision) for each initial state (before collision). The necessary memory can be estimated as follows. The coding of a single state requires 24 bits. On a CRAY two states can be packed into one word (8 bytes = 64 bits). Thus for the storage of $2^{24} = 16\ 777\ 216$ initial and the same number of final states 8 times 16 777 216 bytes or approximately 130 Mbytes are required. This is a severe obstacle for computers with small memories.

Relatively few applications of the FCHC model have been published. A paper of Chen et al. (1991c) on the flow through porous media is especially remarkable. The calculations were performed on a CRAY-YMP (core size of a few hundred Mbytes). The size of the look-up table which actually should contain 16 million entries could be somewhat reduced by exploiting the hole-particle symmetry. The propagation has been coded in CRAY-Assembler. Despite all these machine-specific measures the update rates per node are not higher than those of the PI model (compare Section 3.6) in 3D which is coded in standard C (Wolf-Gladrow and Vogeler, 1992). For a fair comparison of the two models, however, the different values of the shear viscosities have to be taken into account.

The collision rules were simplified quite drastically by Rem and Somers (1989) even risking the violation of the semi-detailed balance. The resulting look-up table requires only 40 kbytes. Computer experiments show good agreement with theoretical predictions such as Fermi-Dirac distribution and the shear viscosity as a function of density. Yet, the shear viscosity is three time higher than its optimal value.

In addition to his isometric collision rule Hénon has proposed a purely random rule: The final state will be randomly chosen out of all states with the same mass and momentum as the initial state. The values of the shear viscosity for both rules proposed by Hénon are comparable (Hénon, 1987a).

3.5.3 Isometric rules for HPP and FHP

Hénon applied his method for the construction of isometric collision rules also to the HPP and the FHP model (without restparticles). For HPP one obtains collision rules which are identical to those of the original formulation: the head-on collision is the only collision. The model is too simple to allow additional collisions. On the other hand the Hénon constraints are not too restrictive to forbit all nontrivial transformations.

For the FHP model the isometric collision rules read

$$(i, i+3) \rightarrow (i+1, i+4) \text{ or } (i-1, i+2),$$
$$(i, i+2, i+4) \rightarrow (i+1, i+3, i+5) \text{ or } (i, i+2, i+4),$$
$$(i-1, i, i+2) \rightarrow (i-2, i, i+1),$$
$$(i+1, i+2, i+4, i+5) \rightarrow (i, i+2, i+3, i+5) \text{ or }$$
$$(i, i+1, i+3, i+4),$$

i.e. the two-particle, the three-particle, the two-particle with observer and the four-particle collisions. The only difference in relation to the original formulation (compare Section 3.2) occurs for the three-particle collision: according to the isometric rules the particle velocities will be changed with a probability of 1/2 instead of 1 as in the original rules. This can be interpreted as an indication that the three-particle collision does not contribute much to the reduction of the shear viscosity. This collision was introduced to destroy a spurious invariant.

3.5.4 What else?

- Implementation of the FCHC model: in the 4th dimension the model encompasses only two 'layers' and periodic boundary condition.
- The fourth component of the momentum behaves like a passive scalar (Frisch et al., 1987).
- Variants of the FCHC model are listed in Table 3.5.1.
- Further reading: Rivet et al. (1988), Shimomura et al. (1988), Hénon (1989), Cancelliere et al. (1990), Dubrulle et al. (1990), Ladd and Frenkel (1990), Vergassola et al. (1990), Rivet (1991), Benzi et al. (1992), Hénon (1992), van der Hoef et al. (1992), Verheggen (1992), van Coevorden et al. (1994), Adler et al. (1995).

Exercise 3.5.1. (**)
Classify the 64 different states at each node of the FHP-I model according to particle number and momentum.

Exercise 3.5.2. (**)
How many different 3-momenta can be realized in the FCHC model?

Concluding remark: If one restricts oneself to models with a single lattice speed there exists only the FCHC model for hydrodynamic simulations in 3D. The collision rules of the FCHC model are much more complicated than those of the FHP model. Later on we will discuss multi-speed models as an alternative to FCHC.

Table 3.5.1. *Variants of the FCHC model (compare Dubrulle et al., 1990). Roughly, the Reynolds coefficient R_*^{max} measures the inverse viscosity in lattice units.*

Name	Rest particles	Semi-detailed balance	R_*^{max} Boltzmann	R_*^{max} measured	References
FCHC-1	0	Yes	2.00	2.0	Hénon (1987), Rivet (1987)
FCHC-2	0	Yes	6.44	-	Hénon (1989)
FCHC-3	0	Yes	7.13	6.4	Hénon (1989), Rivet (1988a,b)
FCHC-4	0	Yes	7.57	-	Hénon (1989)
FCHC-5	3	Yes	10.71	-	Hénon (1989)
FCHC-6	0	No	17.2	-	Dubrulle (1988)
FCHC-7	3	No	(∞)	7.9	Dubrulle et al. (1990)
FCHC-8	3	No	99.7	13.5	Dubrulle et al. (1990)

3.6 The pair interaction (PI) lattice-gas cellular automata

The FCHC model for hydrodynamic simulations in 3D has complicated colli-
sion rules and thus requires large look-up tables. Extension to problems with
a free surface or to magneto-hydrodynamics seems to be extremely involved.
In 1989 Nasilowski proposed a lattice-gas cellular automata which runs in
2D over a square lattice as well as in 3D over a cubic grid. In contrast to
FHP or FCHC the state of a cell of the PI model is characterized by $D+1$
(D spatial dimension) bits instead of only one. The interaction[37] at a node
consists of a succession of interactions between pairs of cells (thus the name
PI = pair interaction). The splitting into pair interactions allows an efficient
coding with bit-operators (C) or bit-functions (FORTRAN) also in 3D.

A complete discussion of the pair interaction lattice-gas cellular automata
has been given by Nasilowski (1991). Here only the main ideas of the PI-
approach will be explained and consequentely the presentation is restricted
to the two-dimensional case. The extension to 3D is straightforward.

3.6.1 Lattice, cells, and interaction in 2D

The PI model in 2D is based on the square lattice (compare Fig. 3.6.1). As
usual the development in time proceeds by an alternating sequence of local
interaction (only cells of a single node are involved) and propagation to the
nearest neighbor nodes. The lattice splits into two sub-lattices[38]:

- At even time levels the particles reside on nodes with even indices (white
  circles).
- At odd time levels the particles reside on nodes with odd indices (black
  circles).

As for HPP, FHP and FCHC there is a cell on each link at each node. The
state of a cell of the PI model is characterized by $D+1$ bits n_J ($J = 0, 1, ..., D$)
where D is the dimension: n_0 is called the *mass bit* and n_j ($j = 1, ..., D$) are
the *momentum bits*. By convention in this section uppercase indices run from
0 to D whereas lowercase indices run from 1 to D. The momentum bits are
subject to the constraint

---

[37] We speak of interaction instead of collision because some of the rules of the PI
lattice-gas cellular automata cannot be described as collisions between particles.

[38] Points with a combination of an even and an odd index will never be occupied
by particles. Therefore they will not be called nodes and they are not shown in
Fig. 3.6.1.

$$n_j \leq n_0 \tag{3.6.1}$$

which can be interpreted as 'the momentum of empty cells vanishes'. The vector linking two neighboring nodes is termed c (lattice velocity). The components of c take on the values 1 or -1. The links to the neighboring nodes and the corresponding cells of a node will be labelled a, b, c, d (compare Fig. 3.6.2). The corresponding lattice velocities are given by $c_a = (1,1)$, $c_b = (-1,1)$, $c_c = (-1,-1)$, $c_d = (1,-1)$. The *momentum* m is defined *component-wise*:

$$m_{j'} := n_{j'} v_{j'}, \quad j' = 1, ..., D \tag{3.6.2}$$

(remark: no summation convention here!). This definition is rather unusual because in general the momentum does not point to the same direction as the velocity. This can be illustrated by considering all possible states of cell a:

1. Mass bit $n_0 = 0 \rightarrow$ all momentum bits vanish (according to eq. 3.6.2): the cell is empty.

2. Mass bit $n_0 = 1$, all momentum bits vanish: rest particle.

3. $n_0 = n_1 = 1$, $n_2 = 0$: particle with x-momentum only.

4. $n_0 = n_2 = 1$, $n_1 = 0$: particle with y-momentum only.

5. $n_0 = n_1 = n_2 = 1$: particle with momentum in the diagonal direction; this is the only case where m and c point to the same direction.

Fig. 3.6.1. *The sub-lattices of the PI lattice-gas cellular automata in 2D*

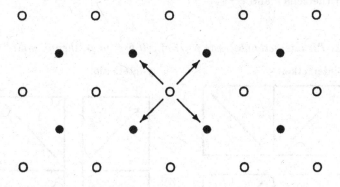

What was Nasilowski's motivation to introduce the somewhat strange definition of the momentum? There are at least two good reasons:

1. The component-wise definition of the momentum allows a splitting of the interaction into pair interactions (see below).

Fig. 3.6.2. *Structure of the nodes and cells of the PI lattice-gas cellular automata.*

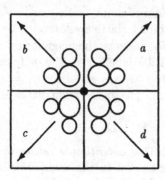 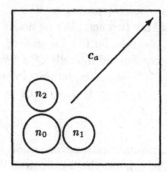

2. The symmetry of the square lattice is not sufficient to assure the isotropy of 4th rank lattice tensors (compare HPP and Section 3.3 on lattice tensors). The component-wise definition of the momentum introduces a *new degree of freedom* in that 'the momentum can fluctuate with respect to the direction of velocity'. This freedom may open a new route to isotropy.

The interaction is composed of the following sequence of pair interactions (compare Fig. 3.6.3):

1. Interaction in x-direction between the cells a and b and between the cells c and d

2. followed by interaction in y-direction between the cells a and d and between the cells b and c.

Fig. 3.6.3. *PI: interaction between pairs of cell first in x-, then in y-direction.*

x-interaction   y-interaction

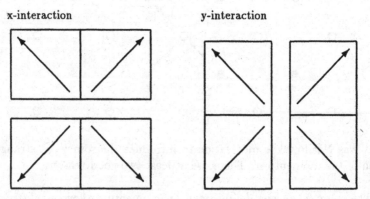

The pair interaction rules are designed according to the maxim 'whatever is not forbidden is allowed'. Nasilowski (1991) formulates three constraints:

1. The interaction must conserve mass and momentum (as all good lattice-gas cellular automata should do).

2. The interaction must be reversible, i.e. the mapping between initial and final state is *one-to-one*. This allows the calculation of the statistical equilibrium by applying the Gibbs formalism.

3. The interaction should yield a maximal change of the state of a node. The identity fulfills the first two constraints but leads to spurious (additional) invariants as, for example, the particle number on each diagonal of the lattice. Such invariants will produce deviations from the hydrodynamic behavior in the macroscopic limit and therefore should be avoided.

The rules given by Nasilowski (1991, p. 107) obey the first and second constraint and encompass all allowed pair interactions (compare also Fig. 3.6.4) which most probably lead to the maximal change.

3.6.2 Macroscopic equations

The rather lengthy calculations of the equilibrium distribution and the multi-scale expansion has been given by Nasilowski (1991). The first order terms of a multi-scale expansion and the rescaling of certain quantities leads for $\rho_0 = 1/2$ to the continuity equation

$$\nabla \cdot u = 0 \tag{3.6.3}$$

and the Euler equation

$$(\partial_t + u\nabla)u + \nabla P = 0 \tag{3.6.4}$$

where

$$P = \frac{p}{\rho_0} = \frac{4}{9}\rho \tag{3.6.5}$$

is the kinematic pressure. The hydrodynamic velocity u is related to the momentum density q by

$$u = \frac{8}{9}q \tag{3.6.6}$$

where the hyper-momentum density q is defined component-wise by

$$q_J = 2^{-D} \sum_v v_J \langle n_{v\,J} \rangle \tag{3.6.7}$$

with the hyper-velocity $v = (v_0, v_1, ..., v_D) = (1, c)$. In particular, $q := (q_1, ..., q_D)$ is the momentum density, and $\rho := q_0$ is the mass density. The viscosity resulting from the second order terms of the multiscale expansion is anisotropic: a tensor of 4th order instead of a scalar.

Fig. 3.6.4. *Pair interaction in horizontal (x-) direction: all possible configurations and changes. Open cycles denote empty cells, filled cycles denote occupied cells, and arrows indicate the momentum.*

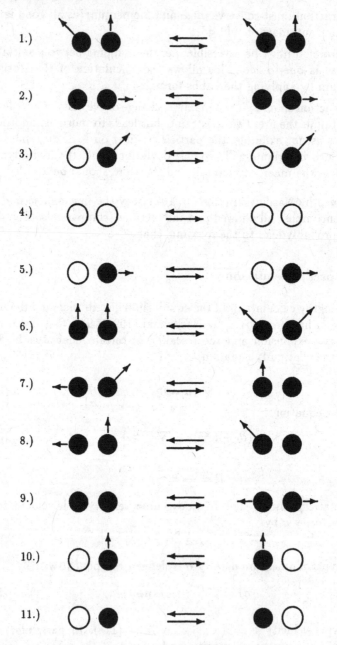

Fig. 3.6.5. *Simulation with PI-LGA of a Karman vortex street in 2D at a Reynolds number of 80 (Wolf-Gladrow et al., 1991): flow past a plate with upstream on the left. The figure shows the perturbation of the velocity after 80,000 time steps. The homogeneous flow field was subtracted to make the eddies clearly visible. The lattice consists of 6400 times 3200 nodes.*

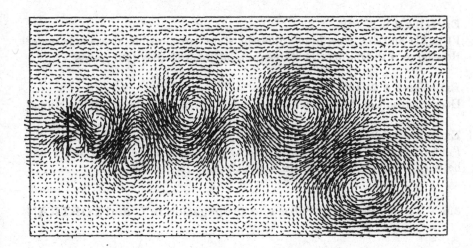

3.6.3 Comparison of PI with FHP and FCHC

The PI-LGCA has some disadvantages compared with FHP and FCHC. PI can only be applied at the particular mass density $\rho_0 = 1/2$ and the viscosity is non-isotropic. On the other hand, the advantages are impressive. The pressure does not depend explicitly on the velocity and simulations in 3D are possible with portable code (Wolf-Gladrow and Vogeler, 1992). As an example Vogeler and Wolf-Gladrow (1993) have calculated drag coefficients for flows past obstacles in two and three dimensions.

Exercise 3.6.1. (*)
PI in 2D and 3D: How many different states are possible at a single node? How large is the number of states on a lattice with only four nodes?

Exercise 3.6.2. (**)
Does the result of the interaction depend on the order of the pair interactions?

Exercise 3.6.3. (**)
Find nontrivial rules for a pair interaction between the cell a and c and between b and d ('*diagonal pair interaction*').

3.6.4 The collision operator and propagation in C and FORTRAN

C:

```
IXM = 32;      IXM1 = IXM - 1;
IYM = 1024;    IYM1 = IYM - 1;

LAST = LENGTH - 1;    /* the last bit */



/* -----   interaction on 1. sub-lattice   -----  */
/* -----   interaction: x-direction         -----  */
/* -----        pair a <--> b               -----  */

for(ix=0; ix < IXM; ix++)
for(iy=1; iy < IYM1; iy++) {

  ab0 = a10[ix][iy] & b10[ix][iy];
  ba0 = a10[ix][iy] ^ b10[ix][iy];
  nab1 = ~(a11[ix][iy] | b11[ix][iy]);
```

```
chab0 = ba0 & nab1;

chab1 = ab0 & ~(a11[ix][iy] ^ b11[ix][iy]);

chab2 = ((ba0 & nab1) | ab0) & (a12[ix][iy] ^ b12[ix][iy]);
/*  array_1 --> array_2 */

  a20[ix][iy] = a10[ix][iy] ^ chab0;
  b20[ix][iy] = b10[ix][iy] ^ chab0;

  a21[ix][iy] = a11[ix][iy] ^ chab1;
  b21[ix][iy] = b11[ix][iy] ^ chab1;

  a22[ix][iy] = a12[ix][iy] ^ chab2;
  b22[ix][iy] = b12[ix][iy] ^ chab2; }

/*  ... interactions c <-> d, a <-> d, b <-> c ...       */

/*        propagation:      */

/* ---    a-direction   --- */

for(ix=0; ix < IXM; ix++)
for(iy=0; iy < IYM; iy++) {
  a10[ix][iy] = a20[ix][iy];
  a11[ix][iy] = a21[ix][iy];
  a12[ix][iy] = a22[ix][iy]; }

/* ---    b-direction   --- */

for(ix=0; ix < IXM1; ix++)
for(iy=0; iy < IYM; iy++) {
 b10[ix][iy] = (b20[ix][iy]>>1) + (b20[ix+1][iy]<<LAST);
 b11[ix][iy] = (b21[ix][iy]>>1) + (b21[ix+1][iy]<<LAST);
 b12[ix][iy] = (b22[ix][iy]>>1) + (b22[ix+1][iy]<<LAST); }
```

FORTRAN:

```
      IXM = 32
      IXM1 = IXM - 1
      IYM = 1024
```

```
      IYM1 = IYM - 1

c     last bit
      LAST = LENGTH - 1

c ----- interaction on 1. sub-lattice -----
c ----- interaction: x-direction    -----


c ------------------ pair (a,b)

      DO iy=1,IYM
      DO ix=1,IXM

      ab0 = iand(a10(ix,iy),b10(ix,iy))
      ba0 = ieor(a10(ix,iy),b10(ix,iy))
      nab1 = not(ior(a11(ix,iy),b11(ix,iy)))

c --- change ?

      chab0 = iand(ba0,iand(nab1,sb1(ix,iy)))
      chab1 = iand(ab0,not(iand(ieor(a11(ix,iy),
     1              b11(ix,iy)),sb1(ix,iy))))
      chab2 = iand(ior(iand(ba0,nab1),ab0),
     1         iand(ieor(a12(ix,iy),b12(ix,iy)),sb1(ix,iy)))

c --- set new values

      a20(ix,iy) = ieor(a10(ix,iy),chab0)
      b20(ix,iy) = ieor(b10(ix,iy),chab0)
      a21(ix,iy) = ieor(a11(ix,iy),chab1)
      b21(ix,iy) = ieor(b11(ix,iy),chab1)
      a22(ix,iy) = ieor(a12(ix,iy),chab2)
      b22(ix,iy) = ieor(b12(ix,iy),chab2)

      ENDDO
      ENDDO

c        ... interactions c <-> d, a <-> d, b <-> c ...

c                    propagation:

c ---   a-direction  ---

      DO iy=1,IYM
```

```
      DO ix=1,IXM
         a10(ix,iy) = a20(ix,iy)
         a11(ix,iy) = a21(ix,iy)
         a12(ix,iy) = a22(ix,iy)
      ENDDO
      ENDDO
c ---    b-direction    ---

      DO iy=1,IYM
      DO ix=1,ixm1
         b10(ix,iy) = ishft(b20(ix,iy),-1)
     1             + ishft(b20(ix+1,iy),LAST)
         b11(ix,iy) = ishft(b21(ix,iy),-1)
     1             + ishft(b21(ix+1,iy),LAST)
         b12(ix,iy) = ishft(b22(ix,iy),-1)
     1             + ishft(b22(ix+1,iy),LAST)
      ENDDO
      ENDDO
```

3.7 Multi-speed and thermal lattice-gas cellular automata

The macroscopic equations derived from the microdynamics of lattice-gas cellular automata include 4th rank tensors which are composed of the lattice velocities and therefore are referred to as lattice tensors (compare Section 3.3 and especially Eq. (3.3.4)). The isotropy of these tensors, which depend on the symmetry of the underlying lattice, is an essential condition to obtain the Navier-Stokes equations. Hasslacher (1987) has shown that models with several different non-vanishing lattice speeds may be equivalent to models with a single speed but larger symmetry group. The resulting 4th rank tensors include speed dependent weights and are referred to as generalized lattice tensors. By appropriate choice of the weights these tensors can be isotropic for relative low symmetry of the lattice. In contrast to single speed models the pressure in multi-speed models does not depend explicitly on the flow velocity. The collision rules can be chosen such that in addition to mass and momentum also kinetic energy will be conserved. Such models are called thermal LGCA.

Models with several different speeds have been encountered before, namely the FHP variants with rest particles (FHP-II, FHP-III). Particles with vanishing speed have no influence on the isotropy of the lattice tensors. Therefore only models with different *non-vanishing* speeds are called *multi-speed models*.

3.7.1 The D3Q19 model

D'Humières, Lallemand, and Frisch proposed already in 1986 - together with FCHC - a multi-speed model in 3D with 19 lattice velocities (Eq. 3.3.24) and three speeds (0, 1, and $\sqrt{2}$). The collisions shall conserve mass ρ, momentum j and kinetic energy density $\rho\varepsilon_K$ which are defined as follows

$$\rho = \sum_I N_I$$

$$j = \rho u = \sum_I N_I c_I$$

$$\rho\varepsilon_K = \sum_I N_I \frac{c_I^2}{2}$$

where the index $I = (\sigma, i)$ indicate the speed ($\sigma = 0, 1, 2$ for the rest particles and particles with speed 1 and $\sqrt{2}$, respectively) as well as the direction i.

The exclusion principle and semi-detailed balance lead to Fermi-Dirac distributions for the mean occupation numbers

$$N_I = f_{FD}(Q_I) = \frac{1}{1 + \exp(Q_I)}$$

whereby

$$Q_I = \alpha + \beta c_I \cdot u + \gamma c_I^2$$

is a linear superposition of the collision invariants. The Lagrange multipliers α, β, and γ can be calculated by expansion for small Mach numbers (compare analogous calculations for FHP).

The D3Q19 model is the first lattice-gas cellular automata with an energy term in the distribution function. This raises the question why similar terms do not occur in the other models (FHP, FCHC, PI).

- For FHP without rest particles mass and kinetic energy density are identical:

$$2\rho\varepsilon_K = \sum_i N_i \underbrace{c_i^2}_{=1} = \sum_i N_i = \rho$$

- FHP with rest particles: the collisions including rest particles do not conserve kinetic energy.

- PI: the kinetic energy density can be defined by

$$\rho\varepsilon_K := \sum_i \frac{p_i^2}{2m}$$

(particle mass $m = 1$); some of the interactions do not conserve energy.

- FCHC: same as for FHP, i.e. the kinetic energy density is proportional to the mass density for the version without rest particles and the collisions including rest particles do not conserve kinetic energy.

The advection term of the macroscopic momentum equation of the D3Q19 model contains the following 4th rank tensor

$$T_{\alpha\beta\gamma\delta} = \frac{\beta_0^2}{2} \sum_I \underbrace{d_\sigma(1 - d_\sigma)(1 - 2d_\sigma)}_{= w_\sigma} c_{I\alpha} c_{I\beta} c_{I\gamma} c_{I\delta}$$

where d_σ are the equilibrium occupation numbers at vanishing flow speed (the zeroth order term of the Taylor expansion) which depend only on the lattice speed and not on direction. The new feature - compared to single speed models - is the occurrence of the weights

$$w_\sigma = d_\sigma(1 - d_\sigma)(1 - 2d_\sigma)$$

which can influence the transformation properties (isotropy) of $T_{\alpha\beta\gamma\delta}$ (of course the coefficient $\beta_0^2/2$ does not play a role). The tensor $T_{\alpha\beta\gamma\delta}$ of the D3Q19 model is isotropic if

$$d_1(1-d_1)(1-2d_1) = 4d_2(1-d_2)(1-2d_2) \qquad (3.7.1)$$

(d'Humières et al., 1986; compare also Section 3.3), i.e. the occupation numbers of speed-1 and speed-$\sqrt{2}$ particles must respect a certain ratio. In other words, the tensor is isotropic only at a certain 'temperature' where the right number of high energy states are excited. For small densities ($d_\sigma \ll 1$) relation (3.7.1) can be approximated by $d_1 = 4d_2$, i.e. the occupation numbers of cells with speed 1 must be four times as high as for cells with speed $\sqrt{2}$ (compare Fig. 3.7.1 for the ratio of d_2/d_1 for finite values of d_1). Thus a certain non-isotropy of the occupation numbers ensures the isotropy of the tensor $T_{\alpha\beta\gamma\delta}$.

Fig. 3.7.1. *The figure shows one solution of the cubic equation* $d_1(1-d_1)(1-2d_1) = 4d_2(1-d_2)(1-2d_2)$ *(solid line). For small values of* d_1 *it can be approximated by* $d_1 = 4d_2$ *(broken line).*

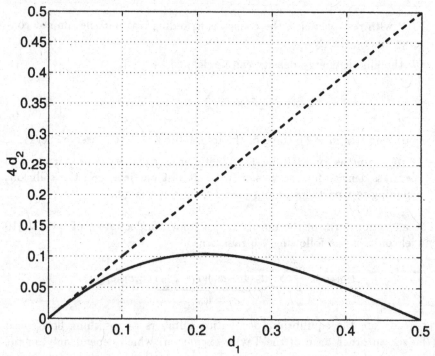

Exercise 3.7.1. (*)
Which interactions of the PI model violate energy conservation?

Exercise 3.7.2. (***)
Propose collision rules for D3Q19.

3.7.2 The D2Q9 model

Chen et al. (1989) reduced the multi-speed model of d'Humières et al. (1986) to 2D. They proposed collision rules and performed some numerical experiments. The model encompasses 9 lattice velocities:

$$
\begin{aligned}
c_0 &= (0,0) & &\text{rest particle} \\
c_{1,2} &= (\pm 1, 0) & &\text{1-particles} \\
c_{3,4} &= (0, \pm 1) & &\text{1-particles} \\
c_{5,6,7,8} &= (\pm 1, \pm 1) & &\sqrt{2}\text{-particles.}
\end{aligned}
$$

All particles have the same mass. Only collisions between two particles are considered (compare Fig. 3.7.2):

1. Head-on collision of two particles with speed $c = 1$: as for HPP both particles are rotated (in the same sense) by $90°$.

2. Head-on collision of two particles with speed $c = \sqrt{2}$: as for HPP both particles are rotated (in the same sense) by $90°$.

3. Collision between a $\sqrt{2}$-particle and a rest particle: two particles with speed $c = 1$ leave the node under $\pm 45°$ with respect to the incoming $\sqrt{2}$-particle. The inverse process is also allowed.

4. Collision between a $\sqrt{2}$-particle and a particle with speed $c = 1$ from different quadrants: the $\sqrt{2}$-particle will be rotated by $90°$ whereas the velocity of the 1-particle will be reversed (identical to rotation by $180°$).

All these two-particle collisions conserve mass, momentum, and kinetic energy. The third type of collisions changes the number of particles with given speed. The tensor of 4th rank in the advection term reads

$$
\begin{aligned}
T_{\alpha\beta\gamma\delta} = {}& const. \cdot (\delta_{\alpha\beta}\delta_{\gamma\delta} + \delta_{\alpha\gamma}\delta_{\beta\delta} + \delta_{\alpha\delta}\delta_{\beta\gamma}) \\
& + 2\left[d_1(1-d_1)(1-2d_1) - 4d_2(1-d_2)(1-2d_2)\right]\delta_{\alpha\beta\gamma\delta}.
\end{aligned}
$$

The second part of the term, which destroys isotropy, vanishes under the same condition (3.7.1) given by d'Humières et al. (1986). In the limit of small occupation numbers one obtains $d_1/d_4 = 4$ and a kinetic energy density $\varepsilon_K = 1/3$. In the same limit the pressure does not depend explicitly on the flow velocity. For finite mass densities, however, there occurs an u^2 term as in FHP.

Furthermore, the model allows simulation of pure heat conduction problems ($u = 0$). However, this is possible with simpler lattice-gas cellular automata (see, for example, Chopard and Droz, 1988 and 1991) or with 'classical' methods like finite differences. Biggs and Humby (1998) summarize their discussion of thermal LGA as follows:

"Thermal LGA has not been applied to problems of any real significance. The only simulation of note is that of Chen et al. (1991a) who briefly considered the Bénard convection problem. The relative paucity of non-isothermal LGA studies is perhaps not surprising given the relative immaturity of this level of LGA model. The models have, however, developed to a point where serious application can be contemplated provided substantial validation work is undertaken."

Exercise 3.7.3. (**)
Show that the collision rules of Chen et al. (1989) contain all possible two-particle collisions.

Exercise 3.7.4. (**)
Find three-particle collisions which conserve mass, momentum, and energy.

Fig. 3.7.2. *Collision rules proposed by Chen et al. (1989) for the D2Q9 model. The rest-particle in rule three is indicated by a circle.*

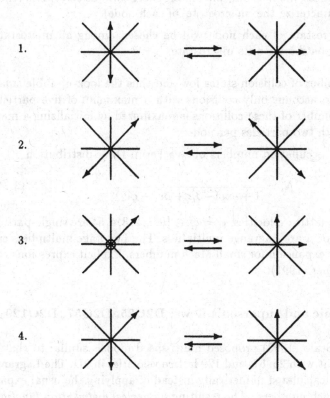

3.7.3 The D2Q21 model

Fahner (1991) proposed a model with 21 lattice velocities in 2D (compare Section 3.3 and especially Eq. (3.3.18)). In addition to the 9 velocities of the D2Q9 model Fahner included the speeds 2 and $\sqrt{5}$. The number of possible states per node of $2^{21} = 2\,097\,152$ is almost comparable with that of FCHC. Fahner proposed the following collision rule which reminds on Hénon's random rule for FCHC:

- At each node mass, momentum, and kinetic energy will be calculated (these quantities characterize the 'macrostate' of each node).
- The final microstate of each node will be chosen among all microstates which are compatible with its macrostate.

To keep the number of collision states low and thus the look-up table small, Fahner took into account only collisions with a maximum of five particles involved. The number of these collisions is maximized by initializing a mean mass density with two particles per node.

The equilibrium occupation numbers obey a Fermi-Dirac distribution

$$N_i = \frac{1}{1 + \exp(-\beta\mu + \beta\epsilon_i - \boldsymbol{q}\boldsymbol{c}_i)} \qquad (3.7.2)$$

where c_i are the lattice velocities, $\epsilon_i = c_i^2 \in \{0, 1, 2, 3, 4, 5\}$ are single-particle energies and $\boldsymbol{q}$, β, μ are Lagrange multipliers. The Lagrange multipliers can be calculated by expansion for small Mach numbers. Explicit expressions can be found in Fahner (1991).

3.7.4 Transsonic and supersonic flows: D2Q25, D2Q57, D2Q129

Kornreich and Scalo (1993) proposed multi-speed models similar to that of Fahner (1991) but with 25, 57, and 129 lattice velocities in 2D. The Lagrange multipliers were calculated numerically instead of applying the usual expansion for small Mach numbers. The resulting *numerical distribution functions* were used to determine the pressure tensor. Kornreich and Scalo found that the velocity dependent term in the expression for the pressure does not increase as fast as it can be predicted by applying the 'expanded' distributions. On the contrary, this term decreases again when the (macroscopic) flow velocity u is near one of the (microscopic) lattice velocities c_i. Thus, in principle simulations of transsonic and supersonic flows are possible. These models are not without problems. The collision rules are complicated already for the D2Q25 model despite the fact that Kornreich and Scalo considered only two-particle collisions. The memory demand is enormous and the viscosity is larger than for FHP.

3.8 Zanetti ('staggered') invariants

In addition to local - at one node, at one time level - invariants there may exist some invariants which involve quantities at different nodes and different time levels. Zanetti (1989, p. 1539) gives a simple example:

"The presence of these new invariants can be easily understood by using a trivial one-dimensional example. Let $g(x)$ be the linear momentum of the particles present at site x, define $G_e(t) = \sum_{x,\text{even}} g(x,t)$, $G_o(t) = \sum_{x,\text{odd}} g(x,t)$ as the total momentum of the particles on even or odd sites, and let the collision rules conserve the momentum and the number of particles at each site. Since the particles can only hop between nearest neighbors, G_e and G_o are exchanged at each time step. The dynamics of this one-dimensional model allows three conserved quantities: M, $G_e + G_o$, and $H = (-1)^t(G_e - G_o)$. The first two are the usual number of particles and the total linear momentum; the third is due to our extremely simplified dynamics."

3.8.1 FHP

Zanetti (1989) has found three[39] staggered invariants of the FHP-III model:

$$H_i = (-1)^t \sum_{r \in \Omega} (-1)^{b_i \cdot r} c_i^{\perp} \cdot j; \quad i = 1, 2, 3, \tag{3.8.1}$$

where $c_i^{\perp}$ is obtained by rotating c_i by $\pi/2$ counterclockwise, $b_i = (2/\sqrt{3})c_i^{\perp}$ is the reciprocal space vector perpendicular to c_i and $j = \sum_k c_k n_k(x,t)$ is the microscopic momentum density.

According to Zanetti the validity of (3.8.1) "can be verified by inspection". More precisely it may be shown (see Exercise 3.8.1) that the H_i are conserved by the combination of any single collision plus propagation and for collisionless propagation. Because the H_i are linear in the n_k this provides a complete proof of the invariance. The Zanetti invariants are also valid for FHP-I and FHP-II which follows as a direct consequence of the proof sketched above.

3.8.2 Significance of the Zanetti invariants

Additional invariants further restrict the dynamics by changing the equilibrium distributions which are functions of all invariants. Fortunately, the procedures usually applied for initialization create very small values of H_k (noise). Consequently, in the simulation reported so far there is no clear indication of the presence of the staggered-momentum density $h_i(r,t) = (-1)^t(-1)^{b_i \cdot r} c_i^{\perp} \cdot j$ $(i = 1,2,3)$. But 'pathological' initial conditions as, for

---

[39] Note: $H_{i+3} = -H_i$ for $i = 1, 2, 3$

example, $\langle \rho \rangle = const$, $\langle j \rangle = 0$, $\langle h_1 \rangle = 0 = \langle h_3 \rangle$ and $\langle h_2 \rangle = h_0 \sin(2\pi y/W)$ lead to an exitation of an oscillatory mode in j_x due to a coupling of the hydrodynamic modes with the additional (kinetic) modes of the lattice gas (Zanetti, 1989).

Further reading: Kadanoff et al. (1989), d'Humières et al. (1989, 1990), Bernardin (1992), Qian (1997). Ernst (1991) calculates the contribution of the unphysical modes to the stress tensor for FHP.

Exercise 3.8.1. (**)
Consider a lattice with 3 times 3 sites which is empty except for the site $(2, 2)$ in the middle of the domain. Calculate the Zanetti invariants 1.) with the initial ($t = 0$) configurations of Fig. 3.2.2 at site $(2, 2)$ and 2.) after collision and propagation ($t = 1$). In addition prove that H_i are conserved by a single propagating particle.

Exercise 3.8.2. (**)
Qian et al. (1992) proposed a LGCA for diffusion in 1D with four velocities c_i (D1Q4) and two different masses m_i:

$$\begin{aligned} c_i &= \{2c, -2c, c, -c\} \\ m_i &= \{m, m, 2m, 2m\} \end{aligned}$$

There is only one possible collision which conserves mass, momentum and energy: the head-on collision between a fast (hot) particle with mass m and a slow (cold) particle with mass $2m$ which leads to a reversal of their velocities. For a domain with L sites where L is divisible by four and periodic boundary conditions find three spurious invariants staggered in space and time and one additional spurious invariant staggered in space only.

Exercise 3.8.3. (**)
Can you find staggered invariants for HPP?

Exercise 3.8.4. (***)
Try to find staggered invariants for PI.

Exercise 3.8.5. (***)
Try to find staggered invariants for D2Q9.

Exercise 3.8.6. (***)
Modify the usual procedure to initialize a given distribution of ρ, j and staggered-momentum densities h_i (the microscopic densities corresponding to H_i).

3.9 Lattice-gas cellular automata: What else?

Further reading:
Textbooks: Rothman and Zaleski (1997), Rivet and Boon (to be published).
Proceedings and reviews:

- Proceedings of the workshop on Large Nonlinear Systems, Complex Systems, Vol. 1, No. 4, 1987.
- Monaco (1989)
- Doolen (1990)
- Ernst (1991)
- Chen et al. (1995)
- Boon et al. (1996)
- Biggs and Humby (1998)

Bibliographies: Doolen (1990) and J. Stat. Phys. 68 (3/4), 611-667, 1992.
Further topics:

- Mode-mode coupling, long time-tails, divergence of transport coefficients in 2D: Kadanoff et al. (1989), Frenkel and Ernst (1989), d'Humières et al. (1989), McNamara (1990), van der Hoef and Frenkel (1990), Ernst (1991).
- Flows in 3D: Bernsdorf et al. (1999), van Genabeek and Rothman (1999).
- Flow through porous media: Balasubramanian et al. (1987), Rothman (1988), Kohring (1991a,b,c), Knackstedt et al. (1993), Gutfraind et al. (1995), van Genabeek and Rothman (1996), Koponen et al. (1996, 1997), Krafczyk et al. (1998), Matsukama (1998), Matsukuma et al. (1998), Niimura (1998), Waite et al. (1998).
- Multiphase flows ('colored models') in 2D: Rothman and Keller (1988), Stockman et al. (1990), Gunstensen and Rothman (1991), Kougias (1993), Rothman and Zaleski (1994, review article), Emerton et al. (1997), Matsukama (1998), Peng and Ohta (1997), Stockman et al. (1997), Tsumaya and Ohashi (1997), Weig et al. (1997), Ebihara et al. (1998), Sehgal et al. (1999).
- Multiphase flows in 3D: Rem and Somers (1989), Olson and Rothman (1995, 1997).
- Flow of granular media: Karolyi et al. (1998), Manna and Khakhar (1998).

- Flow in dynamical geometry: Hasslacher and Meyer (1998)
- Poisson solver: Chen, Matthaeus and Klein (1990).

- Magnetohydrodynamics (MHD): Chen and Matthaeus (1987), Hatori and Montgomery (1987), Montgomery and Doolen (1987), Chen et al. (1991), Succi et al. (1991), Chen et al. (1992), Martinez et al. (1994), Isliker et al. (1998), Takalo et al. (1999).

- Relativistic flows: Balazs et al. (1999).

- Chemical reactions, reaction-diffusion equations: Weimar et al. (1992), Weimar (1997), Decker and Jeulin (1997), Vanag and Nicolis (1999).

- Burgers equation: Boghosian and Levermore (1987), Cheng et al. (1991), Nishinari and Takahashi (1998).

- Generalized semi-detailed balance condition: Chen (1995, 1997).

- Beyond the Boltzmann approximation: Boghosian (1995).

- Pattern formation: Chen et al. (1995), Bussemaker (1996), Deutsch and Lawniczak (1999).

- Integer lattice gases, Digital Physics: Boghosian et al. (1997), Chen, Teixeira and Molvig (1997), Teixeira (1997).

- Quantum mechanics: Boghosian and Taylor (1997), Boghosian and Taylor (1998), Succi (1998), Yepez (1998).

- Further reading: Kohring (1992a,b), Hashimoto and Ohashi (1997), Suarez and Boon (1997), Tribel and Boon (1997), Buick (1998), Hashimoto et al. (1998), Lahaie and Grasso (1998), Masselot and Chopard (1998b), Nicodemi (1998), Tsujimoto and Hirota (1998), Ujita et al. (1998).

4. Some statistical mechanics

4.1 The Boltzmann equation

The motion of a fluid can be described on various levels. The most basic decription by the Hamilton equations for a set of classical particles or the analogous quantum mechanical formulation prohibits itself because of the huge number of particles. 1 cm^3 of air, for example, at 0°C and a pressure of one atmosphere contains $2.69 \cdot 10^{19}$ molecules. It is impossible to prepare a desired microstate of such a system.

This fact is already taken into account by the next higher level of decription for a system with N particles by distribution functions $f_N(q_1, p_1, ..., q_N, p_N, t)$ which encompass all statistical informations of all dynamical processes (q_i and p_i are the generalized coordinate and momentum of particle i). $f_N(q_1, p_1, ..., q_N, p_N, t) dq_1 dp_1 ... dq_N dp_N$ is the probability to find a particle in the interval $([q_1, q_1 + dq_1], [p_1, p_1 + dp_1])$ while the other particles are in infinitesimal intervals around $(q_2, p_2) ... (q_N, p_N)$. Thus f_N contains especially the various correlations between particles. f_N obeys the *Liouville equation*

$$\frac{\partial f_N}{\partial t} - \sum_{j=1}^{3N} \left(\frac{\partial H_N}{\partial q_j} \frac{\partial f_N}{\partial p_j} - \frac{\partial H_N}{\partial p_j} \frac{\partial f_N}{\partial q_j} \right) = 0 \qquad (4.1.1)$$

where H_N is the Hamiltonian of the system.
By integration over part of the phase space one defines *reduced densities*

$$F_s(q_1, p_1, ..., q_s, p_s, t) := V^s \int f_N(q_1, p_1, ..., q_N, p_N, t) \, dq_{s+1} dp_{s+1} ... dq_N dp_N$$

where V^s is a normalization factor. It has been shown that a coupled system of differential equations for the F_s ($1 \leq s \leq N$) is equivalent to the Liouville equation. This system is called *BBGKY* after Bogoljubov, Born, Green, Kirkwood and Yvon who derived these equations. The BBGKY hierarchy has to be truncated at some point to calculate approximate solutions.

The *Boltzmann equation* has been derived as a result of a systematic approximation starting from the BBGKY system not before 1946 (compare Bogoliubov, 1962; Boltzmann derived the equation which bears his name by a different reasoning already in the 19th century). It can be derived by applying the following approximations: 1. Only two-particle collisions are considered (this seems to restict applications to dilute gases). 2. The velocities of the two colliding particles are uncorrelated before collision. This assumption is often called the *molecular chaos hypothesis*. 3. External forces do not influence the local collision dynamics.

The Boltzmann equation is an integro-differential equation for the single particle distribution function $f(x, v, t) \propto F_1(q_1, p_1, t)$

$$\partial_t f + v \partial_x f + \frac{K}{m} \partial_v f = Q(f, f) \qquad (4.1.2)$$

where $x = q_1$, $v = p_1/m$, $m = const$ is the particle mass, $f(x, v, t) d^3x \, d^3v$ is the probability to find a particle in the volume d^3x around x and with velocity between v and $v + dv$.

$$Q(f, f) = \int d^3v_1 \int d\Omega \, \sigma(\Omega) |v - v_1| [f(v')f(v'_1) - f(v)f(v_1)] \qquad (4.1.3)$$

is the *collision integral* with $\sigma(\Omega)$ the *differential collision cross section* for the two-particle collision which transforms the velocities from $\{v, v_1\}$ (incoming) into $\{v', v'_1\}$ (outgoing). K is the body force. It will be neglected in the following discussion of the current chapter.

4.1.1 Five collision invariants and Maxwell's distribution

It can be shown (see, for example, Cercignani, 1988) that the collision integral possesses exactly five elementary *collision invariants* $\psi_k(v)$ $(k = 0, 1, 2, 3, 4)$ in the sense that

$$\int Q(f, f)\psi_k(v) \, d^3v = 0. \qquad (4.1.4)$$

The elementary collision invariants read $\psi_0 = 1$, $(\psi_1, \psi_2, \psi_3) = v$ and $\psi_4 = v^2$ (proportional to mass, momentum and kinetic energy). General collision invariants $\phi(v)$ can be written as linear combinations of the ψ_k

$$\phi(v) = a + b \cdot v + cv^2.$$

It can be further shown (see, for example, Cercignani, 1988) that positive functions f exist which give a vanishing collision integral

$$Q(f, f) = 0.$$

These functions are all of the form

$$f(v) = exp(a + b \cdot v + cv^2)$$

where c must be negative. The *Maxwell*[1] *distribution*

$$f^{(M)} = f(\boldsymbol{x}, \boldsymbol{v}, t) = n \left(\frac{m}{2\pi k_B T} \right)^{3/2} \exp\left[-\frac{m}{2k_B T}(\boldsymbol{v} - \boldsymbol{u})^2 \right] \qquad (4.1.5)$$

is a special case among these solutions where $\boldsymbol{u}$ is the mean velocity

$$\boldsymbol{u} = \frac{1}{n} \int d^3 v \, \boldsymbol{v} f(\boldsymbol{x}, \boldsymbol{v}, t). \qquad (4.1.6)$$

Please note that $f^{(M)}$ depends on $\boldsymbol{x}$ only implicitly via $n(\boldsymbol{x})$, $\boldsymbol{u}(\boldsymbol{x})$ and $T(\boldsymbol{x})$.

4.1.2 Boltzmann's H-theorem

In 1872 Boltzmann showed that the quantity

$$H(t) := \int d^3 v \, d^3 x \, f(\boldsymbol{x}, \boldsymbol{v}, t) \ln f(\boldsymbol{x}, \boldsymbol{v}, t) \qquad (4.1.7)$$

where $f(\boldsymbol{x}, \boldsymbol{v}, t)$ is any function that satisfies the Boltzmann equation fulfills the equation.

$$\frac{dH}{dt} \le 0 \qquad (4.1.8)$$

and the equal sign applies only if f is a Maxwell distribution (4.1.5). This is his famous *H-theorem*.
Proof. We will assume that no external forces are applied and thus $f(\boldsymbol{x}, \boldsymbol{v}, t)$ obeys the following Botzmann equation

$$\frac{\partial f}{\partial t} + \boldsymbol{v} \cdot \boldsymbol{\nabla} f = Q(f, f).$$

Differentiation of (4.1.7) yields

$$\frac{dH}{dt} = \int d^3 v \, d^3 x \, \frac{\partial f}{\partial t}(\boldsymbol{x}, \boldsymbol{v}, t)\left[1 + \ln f(\boldsymbol{x}, \boldsymbol{v}, t)\right].$$

Insertion of the Boltzmann equation leads to

$$\frac{dH}{dt} = -\int \boldsymbol{v} \cdot \boldsymbol{\nabla} \left[f(\boldsymbol{x}, \boldsymbol{v}, t) \ln f(\boldsymbol{x}, \boldsymbol{v}, t) \right] d^3 v \, d^3 x$$

$$\qquad (4.1.9)$$

$$+ \int d^3 v_1 \, d^3 v_2 \, d^3 x \, d\Omega \, \sigma(\Omega) \left[f_2' f_1' - f_2 f_1 \right] \cdot |\, v_2 - v_1 \,| \left[1 + \ln f_1\right]$$

---

[1] The *Maxwell distribution* is also referred to as *Maxwell-Boltzmann distribution* or as *Boltzmann distribution*.

where $f_1 = f(x, v_1, t)$, $f_2 = f(x, v_2, t)$, $f_1' = f(x, v_1', t)$, and $f_2' = f(x, v_2', t)$. The first summand can be transformed into a surface integral

$$- \int_F n \cdot v f(x, v, t) \ln f(x, v, t) \, d^3 v \, dF \qquad (4.1.10)$$

where n is the (outer) normal of the surface F that enclosed the gas. Without detailed discussion we will assume that this surface integral vanishes. The second integral is invariant under exchange of v_1 and v_2 because $\sigma(\Omega)$ is invariant under such exchange:

$$\frac{dH}{dt} = \int d^3 v_1 \, d^3 v_2 \, d^3 x \, d\Omega \, \sigma(\Omega) \mid v_2 - v_1 \mid (f_2' f_1' - f_2 f_1)[1 + \ln f_2] \quad (4.1.11)$$

Adding up half of (4.1.9) and half of (4.1.11) leads to

$$\frac{dH}{dt} = \frac{1}{2} \int d^3 v_1 \, d^3 v_2 \, d^3 x \, d\Omega \, \sigma(\Omega) \mid v_2 - v_1 \mid (f_2' f_1' - f_2 f_1)[2 + \ln(f_1 f_2)]$$
$$(4.1.12)$$

This integral is invariant under exchange of $\{v_1, v_2\}$ and $\{v_1', v_2'\}$ because for each collision there exists an inverse collision with the same cross section. Therefore, one obtains

$$\frac{dH}{dt} = \frac{1}{2} \int d^3 v_1' \, d^3 v_2' \, d^3 x \, d\Omega \, \sigma'(\Omega) \mid v_2' - v_1' \mid (f_2 f_1 - f_2' f_1')[2 + \ln(f_1' f_2')]$$

and because of $d^3 v_1' d^3 v_2' = d^3 v_1 d^3 v_2$, $\mid v_2' - v_1' \mid = \mid v_2 - v_1 \mid$ and $\sigma'(\Omega) = \sigma(\Omega)$:

$$\frac{dH}{dt} = \frac{1}{2} \int d^3 v_1 \, d^3 v_2 \, d^3 x \, d\Omega \, \sigma(\Omega) \mid v_2 - v_1 \mid (f_2 f_1 - f_2' f_1')[2 + \ln(f_1' f_2')].$$
$$(4.1.13)$$

Adding up half of (4.1.12) and half of (4.1.13) leads to

$$\frac{dH}{dt} = \frac{1}{2} \int d^3 v_1 \, d^3 v_2 \, d^3 x \, d\Omega \, \sigma(\Omega) \mid v_2 - v_1 \mid (f_2' f_1' - f_2 f_1)[\ln(f_1 f_2) - \ln(f_1' f_2')].$$

The integrand is never positive because of the inequality

$$(b - a) \cdot (\ln a - \ln b) > 0, \quad a \neq b > 0,$$

thus $dH/dt \leq 0$.

It vanishes, however, when $(f_2' f_1' - f_2 f_1) = 0$ and therefore $\dfrac{\partial f(v, t)}{\partial t} = 0$. $dH/dt = 0$ is possible if and only if

$$f(v_1') f(v_2') - f(v_1) f(v_2) = 0 \qquad (4.1.14)$$

for all v_1', v_2' that result from v_1, v_2 by collisions. From (4.1.14) one obtains

$$\ln f(v_1') + \ln f(v_2') = \ln f(v_1) + \ln f(v_2), \qquad (4.1.15)$$

i.e. $\ln f(v)$ is an additive collision invariant and thus it is of the form (linear composition of the five collision invariants):

$$\ln f(x, v) = a(x) + b(x) \cdot v + c(x)v^2. \tag{4.1.16}$$

Therefore it follows that

$$f(x, v) = C(x)e^{-\dfrac{m(v - u(x))^2}{2k_B T(x)}} \tag{4.1.17}$$

where $C(x)$, $u(x)$ and $T(x)$ are independent of v. However, the distribution (4.1.17) represents no equilibrium state because if $f(x, v, t_1)$ at time t_1 is of the form (4.1.17) then it follows from the Boltzmann equation that

$$\left(\frac{\partial f(x, v, t)}{\partial t}\right)_{t=t_1} = -v \cdot \nabla_x \left[C(x)e^{-\dfrac{m(v - u(x))^2}{2T(x)}}\right] \tag{4.1.18}$$

(the collision term $Q(f, f)$ vanishes because f(x,v) is a function of collision invariants). For the equilibrium state f must be of the form (4.1.17) and be independent of x thus

$$f^{(eq)}(x, v) = f^{(eq)}(v) = Ce^{-\dfrac{m(v - u)^2}{2k_B T}} \tag{4.1.19}$$

with *constants* C, T, u. In a closed system at rest the mean velocity u must vanish and therefore

$$f^{(eq)}(v) = Ce^{-\dfrac{mv^2}{2k_B T}}. \tag{4.1.20}$$

This is the famous Maxwell velocity distribution. **q.e.d.**

4.1.3 The BGK approximation

One of the major problems when dealing with the Boltzmann equation is the complicated nature of the collision integral. It is therefore not surprising that alternative, simpler expressions have been proposed. The idea behind this replacement is that the large amount of detail of two-body interactions is not likely to influence significantly the values of many experimentally measured quantities (Cercignani, 1990).

The simpler operator $J(f)$ which replaces the collision operator $Q(f, f)$ should respect two constraints:

1. $J(f)$ conserves the collision invariants ψ_k of $Q(f, f)$, that is

$$\int \psi_k J(f)\, d^3x\, d^3v = 0 \quad (k = 0, 1, 2, 3, 4), \tag{4.1.21}$$

2. The collision term expresses the tendency to a Maxwellian distribution (H-theorem).

Both constraints are fulfilled by the most widely known model called usually the BGK approximation. It was proposed by Bhatnagar, Gross and Krook (1954) and independently at about the same time by Welander (1954). The simplest way to take the second constraint into account is to imagine that each collision changes the distribution function $f(\boldsymbol{x}, \boldsymbol{v})$ by an amount proportional to the departure of f from a Maxwellian $f^M(\boldsymbol{x}, \boldsymbol{v})$:

$$J(f) = \omega \left[f^M(\boldsymbol{x}, \boldsymbol{v}) - f(\boldsymbol{x}, \boldsymbol{v}) \right].\tag{4.1.22}$$

The coefficient ω is called the collision frequency. From the first constraint it follows

$$\int \psi_k J(f)\, d^3x\, d^3v = \omega \left[\int\int \psi_k f^M(\boldsymbol{x}, \boldsymbol{v})\, d^3x\, d^3v - \int \psi_k f(\boldsymbol{x}, \boldsymbol{v})\, d^3x\, d^3v \right] = 0$$
$$\tag{4.1.23}$$

i.e. at any space point and time instant the Maxwellian $f^M(\boldsymbol{x}, \boldsymbol{v})$ must have exactly the same density, velocity and temperature of the gas as given by the distribution $f(\boldsymbol{x}, \boldsymbol{v})$. Since these values will in general vary with space and time $f^M(\boldsymbol{x}, \boldsymbol{v})$ is called the *local Maxwellian*. Other *model equations* are discussed in Cercignani (1990).

4.2 Chapman-Enskog: From Boltzmann to Navier-Stokes

Many fluid-dynamical phenomena including laminar flows, turbulence and solitons can be described by solutions of the Navier-Stokes equation. Although the form of this equation can be obtained by phenomenological reasoning (see, for example, Landau and Lifshitz, 1959) it is of fundamental as well as practical interest to derive the Navier-Stokes equation (Eq. 1.3.1) from the Boltzmann equation. Applying certain models of the microscopic collision processes one can obtain explicit formulas for the transport coefficients. For example, Maxwell was able to derive an analytical expression for the shear viscosity for molecules which interact by a r^{-5}-potential where r is their distance. It came as a surprise for him and his contemporaries that this theory predicted a dynamic viscosity coefficient independent of density. Experiments made thereafter indeed showed that this is a good approximation for gases over a wide range of densities.

The derivation of the Navier-Stokes equation and its transport coefficients from the Boltzmann equation and certain microscopic collision models runs under the name Chapman-Enskog expansion. This method has been developed by Chapman and Enskog between 1910 and 1920 (Chapman, 1916 and 1918; Enskog, 1917 and 1922; see also Cercignani, 1988 and 1990). The calculations for certain models are rather involved and may easily hide some peculiarities of this expansion. Therefore it seems appropriate to discuss a few interesting features before beginning with the formal derivations and to restrict the calculation to a simple collision model, namely the BGK approximation.

The Chapman-Enskog or multi-scale expansion has already been used to derive the Euler equation for the FHP lattice-gas cellular automata (compare Section 3.2) and will be applied later on to derive the Navier-Stokes and other macroscopic equations for lattice Boltzmann models (compare Section 5.2).

The transformation from the Boltzmann equation to the Navier-Stokes equation involves a *contraction of the description* of the temporal development of the system (Uhlenbeck and Ford, 1963). Whereas the distribution function f of the Boltzmann equation in general is *explicitly* depending on time, space and velocity, we will see that the distribution functions $f^{(n)}$ of the Chapman-Enskog expansion depend only *implicitly* on time via the local density, velocity and temperature, i.e. the $f^{(n)}$ are not the most general solutions of the Boltzmann equation. It can be shown that arbitrary initial distributions relax very fast (a few collision time scales which means of the order of 10^{-11} s in a gas in 3D under standard conditions) toward this special kind of distribution. The possibility of the contraction of the description has been considered as a *very fundamental insight* (Uhlenbeck and Ford, 1963).

The expansion parameter of Chapman-Enskog is the *Knudsen number* K_n, i.e. the ratio between the *mean free length* λ (the mean distance between two succesive collisions) and the characteristic spatial scale of the system (for example, radius of an obstacle in a flow). When the Knudsen number is of the order of 1 or larger the gas in the system under consideration cannot be described as a fluid.

As a last point one should mention that the series resulting from the Chapman-Enskog procedure is probably not convergent but asymptotic[2]. This is suggested by the application to the dispersion of sound (Uhlenbeck and Ford, 1963). Higher order approximations of the Chapman-Enskog method lead to the Burnett and super-Burnett equations (Burnett, 1935, 1936) which have never been applied systematically. One of the problems with these equations is the question of appropriate boundary conditions (see, for example, Cercignani, 1988 and 1990, for further discussion).

4.2.1 The conservation laws

Conservation laws can be obtained by multiplying the Boltzmann equation with a collision invariant $\psi_k(v)$ ($\psi_0 = 1$, $\psi_\alpha = u_\alpha$ for $\alpha = 1, 2, 3$ and $\psi_4' = \frac{1}{2}m|v - u|^2$) and subsequent integration over d^3v. The integrals over the collision integral $Q(f, f)$ vanish by definition. Therefore

$$\int d^3v \, \psi_k (\partial_t + v_\alpha \partial_{x_\alpha}) f(x, v, t) = 0 \qquad (4.2.1)$$

and thus (in 3D)

$$\partial_t \rho + \partial_{x_\alpha}(\rho u_\alpha) = 0 \qquad (4.2.2)$$

$$\rho \partial_t u_\alpha + \rho u_\beta \partial_{x_\beta} u_\alpha = -\partial_{x_\alpha} \hat{P}_{\alpha\beta} \qquad (4.2.3)$$

$$\rho \partial_t \theta + \rho u_\beta \partial_{x_\beta} \theta = -\frac{2}{3} \partial_{x_\alpha} q_\alpha - \frac{2}{3} \hat{P}_{\alpha\beta} \Lambda_{\alpha\beta} \qquad (4.2.4)$$

with

$$n(x, t) = \int d^3v \, f(x, v, t) \qquad (4.2.5)$$

$$\rho(x, t) = m \, n(x, t) \quad (m = const) \qquad (4.2.6)$$

$$\qquad (4.2.7)$$

$$\rho u_\alpha(x, t) = m \int d^3v \, v_\alpha \, f(x, v, t) \qquad (4.2.8)$$

---

[2] Asymptotic series are discussed, for example, in Bender and Orszag (1978). Despite their missing convergence these series can be extremely useful. Bender and Orszag give a number of neat examples.

$$\theta(x,t) = k_B T(x,t) = \frac{m}{3n} \int d^3v \, (v_\alpha - u_\alpha)(v_\alpha - u_\alpha) f(x,v,t)$$

$$\tag{4.2.9}$$

$$\Lambda_{\alpha\beta} = \frac{m}{2}(\partial_{x_\beta} u_\alpha + \partial_{x_\alpha} u_\beta) \tag{4.2.10}$$

$$\hat{P}_{\alpha\beta} = m \int d^3v \, (v_\alpha - u_\alpha)(v_\beta - u_\beta) f(x,v,t) \tag{4.2.11}$$

$$q_\alpha(x,t) = \frac{m^2}{2} \int d^3v \, (v_\alpha - u_\alpha)(v_\beta - u_\beta)(v_\beta - u_\beta) f(x,v,t)$$

$$\tag{4.2.12}$$

Although the conservation equations are exact they are useless until one can solve the Boltzmann equation and apply the solution f to calculate (4.2.5) to (4.2.12). Please note that $\hat{P}_{\alpha\beta}$ is different from the momentum flux tensor introduced in Eq. (3.2.54) in that it does not contain the advection term.

4.2.2 The Euler equation

Inserting $f^{(0)} = f^{(M)}$ (the Maxwell distribution, compare Eq. 4.1.5) into Eqs. (4.2.5) to (4.2.12) leads to the following approximation of the conservation laws

$$\partial_t \rho + \partial_{x_\alpha}(\rho u_\alpha) = 0 \quad \text{(continuity equation)}$$

$$\rho \partial_t u_\alpha + \rho u_\beta \partial_{x_\beta} u_\alpha = -\partial_{x_\alpha} p \quad \text{(Euler equation)}$$

$$\partial_t \theta + u_\beta \partial_{x_\beta} \theta = -\frac{1}{c_v} \theta \, \partial_{x_\alpha} u_\alpha$$

where $p = nk_B T = n\theta$ is the pressure and $c_v = 3/2$ is the heat capacity at constant volume. The heat flux q vanishes in this approximation. The continuity equation is already in its final form. The dissipative terms in the equation of motion have to be derived from higher order approximation.

4.2.3 Chapman-Enskog expansion

The distribution function is expanded as follows

$$f = f^{(0)} + \epsilon f^{(1)} + \epsilon^2 f^{(2)} + \dots \tag{4.2.13}$$

The symbol ϵ is often used in two different ways:

1. One speaks of an expansion as a power series in the small quantity ϵ, i.e. $|\epsilon| \ll 1$. In the case of Chapman-Enskog the Knudsen number K_n can be considered as the small expansion parameter.

2. The formal parameter ϵ in the expansions allows one to keep track of the *relative* orders of magnitude of the various terms. It will be considered only as a *label* and will be dropped out of the final results by setting $\epsilon = 1$.

As an example consider the expansion $f = f^{(0)} + \epsilon f^{(1)}$. In discussions one may consider $f^{(0)}$ and $f^{(1)}$ as quantities of the same order of magnitude and argue that the second term of the expansion is small because ϵ is a small quantity whereas in the formal calculations $f^{(1)}$ is small compared to $f^{(0)}$ and ϵ is only a label to keep track of the relative size of the various terms. The ϵ in this second sense can be set equal to one after finishing all transformations. According to the expansion (4.2.13) the conservation laws (Eqs. 4.2.2 - 4.2.4) can be formulated as follows

$$\partial_t \rho + \partial_{x_\alpha}(\rho u_\alpha) = 0$$

$$\rho \partial_t u_\alpha + \rho u_\beta \partial_{x_\beta} u_\alpha = -\sum_{n=0}^{\infty} \epsilon^n \partial_{x_\alpha} \hat{P}_{\alpha\beta}^{(n)}$$

$$\rho \partial_t \theta + \rho u_\beta \partial_{x_\beta} \theta = -\frac{2}{3} \sum_{n=0}^{\infty} \epsilon^n (\partial_{x_\alpha} q_\alpha^{(n)} + \hat{P}_{\alpha\beta}^{(n)} \Lambda_{\alpha\beta})$$

where

$$\hat{P}_{\alpha\beta}^{(n)} := m \int d^3v\, f^{(n)}(v_\alpha - u_\alpha)(v_\beta - u_\beta) \qquad (4.2.14)$$

and

$$q_\alpha^{(n)} := \frac{m^2}{2} \int d^3v\, f^{(n)}(v_\alpha - u_\alpha)|v - u|^2.$$

Because f depends on t only via ρ, u and T the chain rule

$$\partial_t f = \partial_\rho f\, \partial_t \rho + \partial_{u_\alpha} f\, \partial_t u_\alpha + \partial_\theta f\, \partial_t \theta$$

applies. Inserting (4.2.13) into the derivatives of f with respect to ρ, u_α and T yields

$$\begin{aligned}
\partial_\rho f &= \partial_\rho f^{(0)} &+ \epsilon \partial_\rho f^{(1)} &+ \epsilon^2 \partial_\rho f^{(2)} &+ \dots \\
\partial_{u_\alpha} f &= \partial_{u_\alpha} f^{(0)} &+ \epsilon \partial_{u_\alpha} f^{(1)} &+ \epsilon^2 \partial_{u_\alpha} f^{(2)} &+ \dots \\
\partial_\theta f &= \partial_\theta f^{(0)} &+ \epsilon \partial_\theta f^{(1)} &+ \epsilon^2 \partial_\theta f^{(2)} &+ \dots
\end{aligned}$$

The expansions of $\partial_t \rho$, $\partial_t u_\alpha$ and $\partial_t T$ have to be defined such that they are consistent with the conservation laws in each order of ϵ. The terms of the formal expansion[3]

$$\partial_t = \epsilon \partial_t^{(1)} + \epsilon^2 \partial_t^{(2)} + ... \qquad (4.2.15)$$

will be derived from the conservation laws as follows:

$$\partial_t^{(1)} \rho := -\partial_{x_\alpha}(\rho u_\alpha) \qquad (4.2.16)$$

$$\partial_t^{(n+1)} \rho := 0 \quad (n > 0)$$

$$\partial_t^{(1)} u_\alpha := -u_\beta \partial_{x_\beta} u_\alpha - \frac{1}{\rho} \partial_{x_\beta} \hat{P}_{\alpha\beta}^{(0)} \qquad (4.2.17)$$

$$\partial_t^{(n+1)} u_\alpha := -\frac{1}{\rho} \partial_{x_\beta} \hat{P}_{\alpha\beta}^{(n)} \quad (n > 0)$$

$$\partial_t^{(1)} \theta := -u_\beta \partial_{x_\beta} \theta - \frac{2}{3\rho} \left(\partial_{x_\alpha} q_\alpha^{(0)} + \hat{P}_{\alpha\beta}^{(0)} \Lambda_{\alpha\beta} \right)$$

$$\partial_t^{(n+1)} \theta := -\frac{2}{3\rho} \left(\partial_{x_\alpha} q_\alpha^{(n)} + \hat{P}_{\alpha\beta}^{(n)} \Lambda_{\alpha\beta} \right) \quad (n > 0)$$

Application of these definitions leads to an expansion of $\partial_t f$ into a power series in ϵ:

$$\partial_t f = \left(\epsilon \partial_t^{(1)} + \epsilon^2 \partial_t^{(2)} + ... \right) \left(f^{(0)} + \epsilon f^{(1)} + \epsilon^2 f^{(2)} + ... \right)$$

$$= \epsilon \partial_t^{(1)} f^{(0)} + \epsilon^2 (\partial_t^{(1)} f^{(1)} + \partial_t^{(2)} f^{(0)}) + \epsilon^3 ...$$

Inserting the expansion of the distribution function f into the collision integral $Q(f, f)$ of the Boltzmann equation with BGK approximation[4] yields

$$Q(f, f) = -\omega \left(f - f^{(0)} \right)$$

$$= -\omega \left(\epsilon f^{(1)} + \epsilon^2 f^{(2)} + ... \right)$$

$$=: J^{(0)} + \epsilon J^{(1)} + \epsilon^2 J^{(2)} + ... \qquad (4.2.18)$$

---

[3] The reason for starting the ∂_t expansion by a term linear in ϵ will become apparent from the discussion later on. The expansions of f or ∂_t alone can be multiplied by arbitrary powers of ϵ because the powers of ϵ only label the *relative* size of the different terms in each expansion. When expansions of different quantities are combined, however, the powers of ϵ have to be related such that the terms of leading order yield a meaningful balance.

[4] The BGK approximation will be applied here in order to simplify the calculations.

where

$$J^{(0)}\left(f^{(0)}\right) = 0 \qquad (4.2.19)$$

$$J^{(1)}\left(f^{(0)}, f^{(1)}\right) = J^{(1)}\left(f^{(1)}\right) = -\omega f^{(1)} \qquad (4.2.20)$$

$$J^{(2)}\left(f^{(0)}, f^{(1)}, f^{(2)}\right) = J^{(2)}\left(f^{(2)}\right) = -\omega f^{(2)} \qquad (4.2.21)$$

$$\dots$$

where the collision frequency ω is a constant. In general, i.e. no BGK approximation of the collision integral, the $J^{(n)}$ depend on all $f^{(k)}$ with $k \leq n$ (as indicated on the left hand sides of Eqs. 4.2.20 and 4.2.21) whereas for the BGK approximation $J^{(n)}$ depends only on $f^{(n)}$. This simplification is due to the fact that the collision integral in the BGK approximation is linear in f. The spatial derivative ∂_x on the left hand side of the Boltzmann equation is of the same order as the leading term in the time derivative, i.e.

$$\partial_{x_\alpha} = \epsilon \partial_{x_\alpha}^{(1)}. \qquad (4.2.22)$$

This looks like the first term of an expansion. In space, however, only one macroscopic scale will be considered because different macroscopic processes like advection and diffusion can be distinguished by their time scales but act on similar spatial scales.

Equating terms of same order in ϵ of the Boltzmann equation leads to the following set of equations:

$$J^{(0)}\left(f^{(0)}\right) = 0 \qquad (4.2.23)$$

$$\partial_t^{(1)} f^{(0)} + v_\alpha \partial_{x_\alpha}^{(1)} f^{(0)} = J^{(1)}\left(f^{(0)}, f^{(1)}\right) = -\omega f^{(1)} \qquad (4.2.24)$$

$$\partial_t^{(1)} f^{(1)} + \partial_t^{(2)} f^{(0)} + v_\alpha \partial_{x_\alpha}^{(1)} f^{(1)} = J^{(2)}\left(f^{(0)}, f^{(1)}, f^{(2)}\right) = -\omega f^{(2)}$$

$$\dots$$

Eq. (4.2.23) is fulfilled because J vanishes for Maxwell distributions. $f^{(1)}$ can readily be calculated from Eq. (4.2.24)

$$\boxed{f^{(1)} = -\frac{1}{\omega}\left(\partial_t^{(1)} f^{(0)} + v_\alpha \partial_{x_\alpha}^{(1)} f^{(0)}\right).} \qquad (4.2.25)$$

This equation states that the lowest order deviations $f^{(1)}$ from a local Maxwell distribution $f^{(0)}$ are proportional to the gradient in space and time of $f^{(0)}$. The calculation of $f^{(1)}$ is much more involved when the collision integral is not approximated (see, for example, Huang, 1963).

The next step is the calculation of $\hat{P}^{(1)}_{\alpha\beta}$ according to Eq. (4.2.14)

$$\hat{P}^{(1)}_{\alpha\beta} = m \int d^3v (v_\alpha - u_\alpha)(v_\beta - u_\beta) f^{(1)}$$

$$= -\frac{m}{\omega} \int d^3v (v_\alpha - u_\alpha)(v_\beta - u_\beta)(\partial^{(1)}_t f^{(0)} + v_\gamma \partial^{(1)}_{x_\gamma} f^{(0)}).$$

Insertion of (4.2.16) and (4.2.17) leads to (from now on the superscript [1] will be dropped for the sake of simplicity)

$$\partial_t f^{(0)}(\rho, \boldsymbol{u}) = \frac{\partial f^{(0)}}{\partial \rho}\frac{\partial \rho}{\partial t} + \frac{\partial f^{(0)}}{\partial u_\gamma}\frac{\partial u_\gamma}{\partial t}$$

$$= -\frac{f^{(0)}}{m}\frac{\partial(\rho u_\gamma)}{\partial x_\gamma} + \frac{m}{k_B T}(v_\gamma - u_\gamma)f^{(0)}\left(u_\delta \frac{\partial u_\gamma}{\partial x_\delta} + \frac{1}{\rho}\frac{\partial \hat{P}^{(0)}_{\gamma\delta}}{\partial x_\delta}\right)$$

$$= -\frac{\rho f^{(0)}}{m}\frac{\partial u_\gamma}{\partial x_\gamma} - \frac{f^{(0)}}{m}u_\gamma \frac{\partial \rho}{\partial x_\gamma} + \frac{m}{k_B T}(v_\gamma - u_\gamma)f^{(0)}u_\delta \frac{\partial u_\gamma}{\partial x_\delta}$$

$$+ \frac{m}{k_B T}(v_\gamma - u_\gamma)f^{(0)}\frac{1}{\rho}\delta_{\gamma\delta}\frac{\partial p}{\partial x_\delta}$$

and

$$v_\gamma \partial_{x_\gamma} f^{(0)} = v_\gamma \frac{\partial f^{(0)}}{\partial \rho}\frac{\partial \rho}{\partial x_\gamma} + v_\gamma \frac{\partial f^{(0)}}{\partial u_\delta}\frac{\partial u_\delta}{\partial x_\gamma}$$

$$= v_\gamma \frac{f^{(0)}}{m}\frac{\partial \rho}{\partial x_\gamma} + v_\gamma \frac{m}{k_B T}(v_\delta - u_\delta)f^{(0)}\frac{\partial u_\delta}{\partial x_\gamma}.$$

The various integrals are readily evaluated

$$-\frac{1}{m}\frac{\partial u_\gamma}{\partial x_\gamma}\int d^3v (v_\alpha - u_\alpha)(v_\beta - u_\beta)f^{(0)} = -\delta_{\alpha\beta}\, n \frac{k_B T}{m}\frac{\partial u_\gamma}{\partial x_\gamma}$$

$$\frac{1}{m}\frac{\partial \rho}{\partial x_\gamma}\int d^3v (v_\alpha - u_\alpha)(v_\beta - u_\beta)(v_\gamma - u_\gamma)f^{(0)} = 0$$

$$\frac{m}{k_B T}f^{(0)}\left(u_\delta \frac{\partial u_\gamma}{\partial x_\delta} + \frac{1}{\rho}\delta_{\gamma\delta}\frac{\partial p}{\partial x_\delta}\right)\int d^3v (v_\alpha - u_\alpha)(v_\beta - u_\beta)(v_\gamma - u_\gamma)f^{(0)} = 0$$

$$\frac{m}{k_B T}\frac{\partial u_\delta}{\partial x_\gamma}\int d^3v (v_\alpha - u_\alpha)(v_\beta - u_\beta)v_\gamma(v_\delta - u_\delta)f^{(0)}$$

$$= (\delta_{\alpha\beta}\delta_{\gamma\delta} + \delta_{\alpha\gamma}\delta_{\beta\delta} + \delta_{\alpha\delta}\delta_{\beta\gamma})n\frac{k_B T}{m}\frac{\partial u_\delta}{\partial x_\gamma}$$

and thus

$$
\hat{P}^{(1)}_{\alpha\beta} = -n\frac{k_BT}{\omega}\left[(\delta_{\alpha\beta}\delta_{\gamma\delta} + \delta_{\alpha\gamma}\delta_{\beta\delta} + \delta_{\alpha\delta}\delta_{\beta\gamma})\frac{\partial u_\delta}{\partial x_\gamma} - \delta_{\alpha\beta}\frac{\partial u_\gamma}{\partial x_\gamma}\right]
$$

$$
= n\frac{k_BT}{\omega}\begin{pmatrix} 2\dfrac{\partial u}{\partial x} & \dfrac{\partial u}{\partial y}+\dfrac{\partial v}{\partial x} & \dfrac{\partial u}{\partial z}+\dfrac{\partial w}{\partial x} \\[2mm] \dfrac{\partial u}{\partial y}+\dfrac{\partial v}{\partial x} & 2\dfrac{\partial v}{\partial y} & \dfrac{\partial v}{\partial z}+\dfrac{\partial w}{\partial y} \\[2mm] \dfrac{\partial u}{\partial z}+\dfrac{\partial w}{\partial x} & \dfrac{\partial v}{\partial z}+\dfrac{\partial w}{\partial y} & 2\dfrac{\partial w}{\partial z} \end{pmatrix}.
$$

Neglecting density and temperature variations the divergence of $\hat{P}^{(1)}_{\alpha\beta}$ reads

$$
\frac{\partial \hat{P}^{(1)}_{\alpha\beta}}{\partial x_\alpha} = \mu\left[2\frac{\partial^2 u}{\partial x^2} + \frac{\partial^2 u}{\partial y^2} + \frac{\partial^2 u}{\partial z^2} + \frac{\partial^2 v}{\partial x\partial y} + \frac{\partial^2 w}{\partial x\partial z}\right]e_x + ...
$$

$$
= \mu\left[\frac{\partial}{\partial x_\beta}\left(\frac{\partial u_\alpha}{\partial x_\beta}\right) + \frac{\partial}{\partial x_\alpha}\left(\frac{\partial u_\beta}{\partial x_\beta}\right)\right]
$$

$$
= \mu\left[\nabla^2 u + \nabla(\nabla\cdot u)\right]
$$

where

$$
\mu = n\frac{k_BT}{\omega} \tag{4.2.26}
$$

is the dynamic shear viscosity. Thus one obtains the Navier-Stokes equation

$$
\boxed{\partial_t u_\alpha + u_\beta\partial_{x_\beta} u_\alpha = -\partial_{x_\alpha}P + \nu\partial_{x_\beta}\partial_{x_\beta}u_\alpha + \xi\partial_{x_\alpha}\partial_{x_\beta}u_\beta} \tag{4.2.27}
$$

where the kinematic shear (ν) and bulk (ξ) viscosities are equal and given by

$$
\nu = \frac{k_BT}{\omega m} = \xi. \tag{4.2.28}
$$

4.3 The maximum entropy principle

In 1948 Shannon [418, 419] proposed a theory which allows us to quantify 'information'. The statistical measure of the lack of information is called the information theoretical or Shannon entropy. Equilibrium distributions can be derived from the maximum entropy principle.

The following presentation closely follows Stumpf and Rieckers (1976).

First consider a *discrete set* $Z := \{z_1...z_N\}$ with N elements. A *message*[5] is defined as a selection of one or several elements of Z. The informational measure of the message is defined by that knowledge which is necessary to denote a certain element or a selection of elements. What is the elementary unit of this measure? If the set Z encompasses only one element the selection of this element does not augments our knowledge. There is no real message until the number of elements in Z is at least two. Obviously the decision between two alternatives is the smallest unit of information one can think of: it is called a *bit* which is the short form of '*binary digit*'. The larger the number of elements in Z, the more information is connected with the selection of a certain element of Z. The measure of the information gained can be traced back to a sequence of alternative decisions. The number of elements N can be written down in binary form. The number of binary digits is a measure of information. Or the elements z_j can be arranged in the form of a binary tree (compare Fig. 4.3.1) where the number of branching points from the root to one of the end points equals the number of bits. These procedures work for sets with $N = 2^n$ elements and yield the measure of information $I(N) = n = \log_2 N$ for the selection of a single element. This definition is generalized to sets with arbitrary number of elements by

$$I(N) = \log_2 N,$$

i.e. $I(N)$ is not necessary an integer anymore.

Further the measure of information is additive with respect to the choice of several (α) elements out of a set with N elements

$$I(N, \alpha) = \alpha \cdot I(N)$$

and the choice of two elements out of a direct product of two sets Z_1 and Z_2

---

[5] The notation has its roots in the theory of communication. One of the basic problems in this context is the reliable transmission of messages from a source via a channel to a receiver. Often the messages to be transmitted have meaning like, for example, the news you hear on the radio. This, however, is not always the case. In transmitting music, the meaning is much more subtle then in the case of a verbal message. In any case, meaning is quite irrelevant to the problem of transmitting the information.

Fig. 4.3.1. *The information for the selection of a certain element out of a set of $N = 2^n$ elements is defined as the number of alternative decisions necessary when going from the root to a certain end point. The selection of a certain elements out of 8 elements requires three binary decisions.*

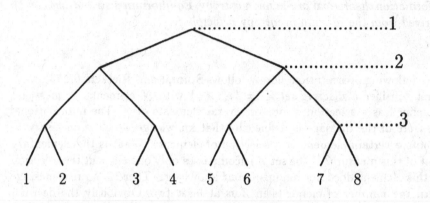

$$I(N_{Z_1 \otimes Z_2}) = I(N_{Z_1}) + I(N_{Z_2}).$$

Now consider *probability distributions* instead of sets. Let us start with discrete probability distributions P with a finite number of entries:

$$P := \{P_1 ... P_N\}, \quad \sum_{k=1}^{N} P_k = 1$$

corresponding to the set of events

$$\Omega := \{x_1 ... x_N\}.$$

The task is to find a measure of information $I(P)$ for a probability distribution as a whole. Let's first consider two special cases:

– The *sharp distribution*: $P_i = \delta_{il}$, i.e. each measurement will yield the result x_l. The sharp distribution contains a lot of information: if you know this probability distribution you can be sure of the output of your next measurement. Because one event out of N possible events is selected the measure

$$I(P) = \log_2 N$$

suggests itself.

- The *normal distribution*: $P_i = 1/N$, i.e. every possible event has the same probability. The normal distribution can be understood as a consequence of the *Laplacian principle of the insufficient reason*: "If there is no reason to single out a certain event for given information concerning an experimental situation, a normal distribution is to be assumed." (Stumpf and Rieckers, 1976, p.14). Obviously the normal distribution contains the minimal amount of information, thus

$$I(P) = 0.$$

The measure of information $I(P)$ for a general distribution $I(P) = I(P_1...P_N)$ is based on four postulates:

1. $I(P)$ is an *universal function*.
2. $I(P)$ is additive concerning the (still to be determined) statistical information for single elements $i(P_k)$ as well as for the composition of direct products:

$$I(P_1...P_N) = \sum_{k=1}^{N} i(P_k)$$

$$I(P_{\Omega_1 \otimes \Omega_2}) = I(P_{\Omega_1}) + I(P_{\Omega_2}).$$

3. $I(P) = 0$ for the normal distribution and $I(P) = \log_2 N$ for the sharp distribution.
4. The statistical information of a single element $i(P)$ is defined on $0 \leq P \leq 1$ and is continuous.

Theorem 4.3.1. *The statistical measure of information $I(P)$ over the set of events Ω with N elements, which fulfills the above given postulates 1 to 4, is uniquely given by*

$$I(P) = I(P_1...P_N) = \sum_{i=1}^{N} P_i \log_2 P_i + \log_2 N. \qquad (4.3.1)$$

The proof can be found, for example, in Stumpf and Rieckers (1976).

Lemma 4.3.1. *The maximum value I_{max} of $I(P)$ is given by the sharp distribution:*

$$I_{max}(P) = \log_2 N.$$

Exercise 4.3.1. (*)
Prove Lemma 4.3.1.

The *information theoretical entropy* or *Shannon entropy* S is defined as follows:

$$S(P_1...P_N) := I_{max} - I(P_1...P_N) = -\sum_{i=1}^{N} P_i \log_2 P_i. \qquad (4.3.2)$$

It is a measure for the *lack of information*: S vanishes for the sharp distribution and becomes maximal for the normal distribution.

The generalization of $I(P)$ for continuous sets of events is given by

$$S[f] := -k \int f(x) \ln f(x) dx, \qquad (4.3.3)$$

i.e. the function $S = S(P)$ is replaced by a functional $S = S[f]$ over the probability density f. The transition from the case of discrete distributions to probability densities f is not as simple as it looks. For example, there is no maximal measure of information and because f can be a generalized function the integral 4.3.3 could be meaningless (see Stumpf and Rieckers, 1976, for a detailed discussion).

The most important theorem of this section reads:

Theorem 4.3.2. *(Maximum entropy principle) If the probability density* $f(x)$ *with the normalization*

$$\int f(x) dx = 1$$

obeys the following m linear independent constraints

$$\int R_i(x) f(x) dx = r_i \quad 1 \leq i \leq m$$

then the probability density which maximizes the lack of information while respecting the $m+1$ constraints is uniquely given by

$$f(x) = \exp\left[-\lambda_0 - \sum_{i=1}^{m} \lambda_i R_i(x)\right] \qquad (4.3.4)$$

where the Lagrange parameters λ_0, λ_1, ..., λ_m *are unique functions of the values* $r_1, ..., r_m$.

Proof. The proof essentially consists of two parts. Here only the derivation of the distribution (4.3.4) shall be discussed in detail. The second part, namely the proof that the Lagrange parameters are uniquely determined by the values $r_1, ..., r_m$, can be found in Stumpf and Rieckers (1976, p. 20) .

The extremum of a functional under given constraints is sought after. An extended functional $\hat{S}[f]$ is defined by coupling the constraints via Lagrange parameters $\eta_0, ...\eta_m$ to the Shannon entropy $S[f]$:

$$\hat{S}[f] \;:=\; S[f] - k(\lambda_0 - 1)\, Tr[f] - k \sum_{i=1}^{m} \lambda_i\, Tr[(R_i - r_i)f]$$

$$= \; -kTr\left[f\left(\ln f + \lambda_0 - 1 + \sum_{i=1}^{m} \lambda_i(R_i - r_i)\right)\right].$$

For reasons which become obvious in a moment the η_j have been written as follows:

$$\eta_0 = k(\lambda_0 - 1)$$
$$\eta_i = k\lambda_i \quad 1 \le i \le m.$$

The *trace* of f is defined as

$$Tr[f] := \int f(x)dx. \tag{4.3.5}$$

From this definition it immediately follows

$$Tr[c \cdot f] = \int c \cdot f(x)\,dx = c \int f(x)\,dx = c\, Tr[f]$$

where c is a constant.

The vanishing of the *functional derivative* of $\hat{S}[f]$ with respect to f is a necessary condition for an extremum of $S[f]$:

$$\frac{\delta \hat{S}[f]}{\delta f} = 0$$

Functional derivative are calculated analogously to the rules for ordinary derivatives (see, for example, Großmann, 1988):

$$\frac{\delta \hat{S}[f]}{\delta f} = -k\left\{\ln f + \lambda_0 + \sum_{i=1}^{m} \lambda_i R_i\right\}$$

and therefore

$$\ln f = -\lambda_0 - \sum_{i=1}^{m} \lambda_i R_i$$

respectively

$$f = \exp\left[-\lambda_0 - \sum_{i=1}^{m} \lambda_i R_i\right].$$

q.e.d.

The maximum entropy principle will be applied later on to calculate equilibrium distributions for lattice Boltzmann models.

Further reading: The proceedings edited by Levine and Tribus (1979) and especially the paper by Jaynes (1979).

Exercise 4.3.2. (**)
Find s_i $(i = 0, 1, ..., l)$ such that

$$\sum_i c_i s_i(x, t) = S(x, t)$$

under the constraint

$$\sum_i s_i(x, t) = 0$$

by minimizing

$$V = \sum s_i^2.$$

The lattice velocities c_i satisfy

$$\sum_i c_i = 0, \quad \text{and} \quad \sum_i c_i^2 = n.$$

Exercise 4.3.3. (**)
The *Renyi entropy* (Renyi, 1970) of order α is definiered as follows:

$$S_\alpha := -\frac{1}{\alpha - 1} \ln \sum_{i=1}^{N} p_i^\alpha, \quad \alpha \in \mathcal{R}, \quad \alpha \neq 1.$$

Calculate

$$\lim_{\alpha \to 1} S_\alpha.$$

5. Lattice Boltzmann Models

5.1 From lattice-gas cellular automata to lattice Boltzmann models

Lattice-gas cellular automata for Navier-Stokes equations are plagued by several *diseases*. Only for some of them therapies and cures could be found (compare Table 5.1.1).

Table 5.1.1. Diseases of lattice-gas cellular automata: its causes and therapies and cures.

disease	cause	therapy/cure	remarks
non-isotropic advection term	lattice tensor of 4th rank is non-isotropic	higher symmetry of lattice / add inner degree of freedom / multi-speed models	HPP → FHP / HPP → PI
violation of the Galilei invariance	Fermi-Dirac distributions	rescaling (symptomatic treatment)	FHP, FCHC, PI
spurious invariants	regular lattices	as much collisions as possible	Zanetti invariants
noise	Boolean variables	averaging (coarse graining)	enormous memory demand
pressure depends explicitly on velocity		multi-speed models	Chen et al. 1989

Historically the following stages in the development of lattice Boltzmann models[1] (LBM) can be distinguished:

---

[1] A lattice Boltzmann model encompasses a lattice, an equilibrium distribution and a kinetic equation which is called lattice Boltzmann equation (LBE). In

1. Lattice Boltzmann equations have been used already at the cradle of lattice-gas cellular automata by Frisch et al. (1987) to calculate the viscosity of LGCA.

2. Lattice Boltzmann models as an independent numerical method for hydrodynamic simulations were introduced by McNamara and Zanetti in 1988. The main motivation for the transition from LGCA to LBM was the desire to get rid of the noise. The Boolean fields were replaced by continuous distributions over the FHP and FCHC lattices. Fermi-Dirac distributions were used as equilibrium functions.

3. Linearized collision operator (Higuera and Jiménez, 1989).

4. Boltzmann instead of Fermi-Dirac distributions.

5. The collision operator, which is based on the collisions of a certain LGCA, has been replaced by the BGK (also called single time relaxation) approximation by Koelman (1991), Qian et al. (1992) and others. These lattice BGK models (*LBGK*) mark a new level of abstraction: collisions are not anymore defined explicitly.

Multi-speed LBGK models are most popular today. However, more complex collision operators are still in use in models of multi-phase flows (Rothman and Zaleski, 1994).

5.1.1 Lattice Boltzmann equation and Boltzmann equation

The microdynamics of LGCA are described by kinetic equations of the type

$$n_i (x + c_i \Delta t, t + \Delta t) = n_i (x, t) + \Delta_i \qquad (5.1.1)$$

(compare, for example, eq. 3.2.15 where Δt was set to 1). Δ_i is the collision function of the respective model. Discrete equations of the form (5.1.1) are referred to as *lattice Boltzmann equations*. The correspondence to the Boltzmann equation

$$\frac{\partial f}{\partial t} + v \nabla f = Q \qquad (5.1.2)$$

(Q is the collision integral; external forces have been neglected) can be shown by expansion of the left hand side of eq. (5.1.1):

$$n_i (x + c_i \Delta t, t + \Delta t) = n_i (x, t) + \Delta t \frac{\partial n_i}{\partial t} + c_i \Delta t \nabla n_i + \mathcal{O}\left((\Delta t)^2\right).$$

Neglecting higher order terms, one obtains

$$\frac{\partial n_i}{\partial t} + c_i \nabla n_i = \frac{\Delta_i}{\Delta t}$$

_____

the literature you will find other names like 'lattice Boltzmann equation' or 'Boltzmann cellular automata' instead of lattice Boltzmann model.

which by substituting $n_i \rightarrow f$, $c_i \rightarrow v$, $\Delta_i/\Delta t \rightarrow Q$ gives the Boltzmann equation (5.1.2).

Sterling and Chen (1996) show that this is more than pure formal correspondence. They derive the lattice Boltzmann equation as a special discretization of the Boltzmann equation. The Boltzmann equation with BGK approximation reads

$$\frac{\partial f}{\partial t} + v\nabla f = -\frac{1}{\tau}\left(f - f^{(eq)}\right).$$

The distribution function f depends on space, velocity and time: $f(x, v, t)$. The v-space is discretized by introducing a finite set of velocities, v_i, and associated distribution functions, $f_i(x, t)$, which are governed by the *discrete Boltzmann equation*:

$$\frac{\partial f_i}{\partial t} + v_i\nabla f_i = -\frac{1}{\tau}\left(f_i - f_i^{(eq)}\right)$$

(please note that this equation is different from the *discretized Boltzmann equation*, see below). The discrete Boltzmann equation will be nondimensionalized by the characteristic length scale, L, the reference speed, U, the reference density, n_r, and the time between particle collisions, t_c,

$$\frac{\partial F_i}{\partial \hat{t}} + c_i\hat{\nabla}F_i = -\frac{1}{\hat{\tau}\epsilon}\left(F_i - F_i^{(eq)}\right) \qquad (5.1.3)$$

where $c_i = v_i/U$, $\hat{\nabla} = L\nabla$, $\hat{t} = t \cdot U/L$, $\hat{\tau} = \tau/t_c$, $F_i = f_i/n_r$. The parameter

$$\epsilon = t_c\frac{U}{L}$$

may be interpreted as either the ratio of collision time to flow time or as the ratio of mean free path to the characteristic length (i.e., Knudsen number). A discretization of eq. (5.1.3) is given by:

$$\frac{F_i\left(\hat{x}, \hat{t} + \Delta\hat{t}\right) - F_i\left(\hat{x}, \hat{t}\right)}{\Delta\hat{t}} + c_{ix}\frac{F_i\left(\hat{x} + \Delta\hat{x}, \hat{t} + \Delta\hat{t}\right) - F_i\left(\hat{x}, \hat{t} + \Delta\hat{t}\right)}{\Delta\hat{x}}$$

$$+c_{iy}\frac{F_i\left(\hat{x} + \Delta y, \hat{t} + \Delta\hat{t}\right) - F_i\left(\hat{x}, \hat{t} + \Delta\hat{t}\right)}{\Delta y}$$

$$+c_{iz}\frac{F_i\left(\hat{x} + \Delta z, \hat{t} + \Delta\hat{t}\right) - F_i\left(\hat{x}, \hat{t} + \Delta\hat{t}\right)}{\Delta z}$$

$$= -\frac{1}{\hat{\tau}\epsilon}\left(F_i - F_i^{(eq)}\right)$$

where $\Delta\hat{t} = \Delta t \cdot U/L$. Lagrangian behavior is then obtained by the selection of the lattice spacing divided by the time step to equal the lattice velocity ($\Delta\hat{x}/\Delta\hat{t} = c_i$):

$$\frac{F_i\left(\hat{x}, \hat{t} + \Delta\hat{t}\right) - F_i\left(\hat{x}, \hat{t}\right)}{\Delta\hat{t}} + \frac{F_i\left(\hat{x} + c_i\Delta\hat{t}, \hat{t} + \Delta\hat{t}\right) - F_i\left(\hat{x}, \hat{t} + \Delta\hat{t}\right)}{\Delta\hat{t}}$$

$$= \frac{F_i\left(\hat{x} + c_i\Delta\hat{t}, \hat{t} + \Delta\hat{t}\right) - F_i\left(\hat{x}, \hat{t}\right)}{\Delta\hat{t}} = -\frac{1}{\hat{\tau}\epsilon}\left(F_i - F_i^{(eq)}\right). \qquad (5.1.4)$$

Thus two terms on the left hand side cancel each other and thereby the method becomes explicit. Choosing $\Delta t = t_c$, multiplying eq. (5.1.4) by $\Delta\hat{t}$ and dropping all carets leads to the *(BGK) lattice Boltzmann equation*

$$\boxed{F_i\left(x + c_i\Delta t, t + \Delta t\right) - F_i\left(x, t\right) = -\frac{1}{\tau}\left(F_i - F_i^{(eq)}\right)} \qquad (5.1.5)$$

Sterling and Chen (1996, p.200) give the following interpretation of Eq. (5.1.5)[2]:

> "This equation has a particular simple physical interpretation in which the collision term is evaluated locally and there is only one streaming step or 'shift' operation per lattice velocity. This stream-and-collide particle interpretation is a result of the fully Lagrangian character of the equation for which the lattice spacing is the distance travelled by the particle during a time step. Higher order discretizations of the discrete Boltzmann equation typically require several 'shift' operations for the evaluation of each derivative and a particle interpretation is less obvious. ... It did not originally occur to the authors" [i.e. McNamara and Zanetti in 1988] "that the LB method could be considered a particular discretization for the discrete Boltzmann equation (G. McNamara, private communication)."

---

[2] Readers not familiar with lattice-gas cellular automata should skip this quotation.

Table 5.1.2. From the Boltzmann equation to the lattice Boltzmann equation.

Boltzmann equation:

$$\frac{\partial f}{\partial t} + v \nabla f = Q$$

Boltzmann equation (BGK approximation):

$$\frac{\partial f}{\partial t} + v \nabla f = -\frac{1}{\tau} \left(f - f^{(eq)} \right)$$

discrete Boltzmann equation:

$$\frac{\partial f_i}{\partial t} + v_i \nabla f_i = -\frac{1}{\tau} \left(f_i - f_i^{(eq)} \right)$$

non-dimensional discrete Boltzmann equation:

$$\frac{\partial F_i}{\partial \hat{t}} + c_i \hat{\nabla} F_i = -\frac{1}{\hat{\tau}\epsilon} \left(F_i - F_i^{(eq)} \right)$$

discretized Boltzmann equation:

$$\frac{F_i \left(\hat{x}, \hat{t} + \Delta \hat{t} \right) - F_i \left(\hat{x}, \hat{t} \right)}{\Delta \hat{t}} + c_{ix} \frac{F_i \left(\hat{x} + \Delta \hat{x}, \hat{t} + \Delta \hat{t} \right) - F_i \left(\hat{x}, \hat{t} + \Delta \hat{t} \right)}{\Delta \hat{x}} \cdots$$
$$= -\frac{1}{\hat{\tau}\epsilon} \left(F_i - F_i^{(eq)} \right)$$

lattice Boltzmann equation:

$$F_i \left(x + c_i \Delta t, t + \Delta t \right) - F_i \left(x, t \right) = -\frac{1}{\tau} \left(F_i - F_i^{(eq)} \right)$$

5.1.2 Lattice Boltzmann models of the first generation

The following remarks are mainly of historical interest and should be skipped by readers not familiar with LGCA.

LGCA are plagued by noise which can be suppressed by coarse graining over large domains and/or time intervals. Thus low noise levels are costly in terms of memory and computer time (see Dahlburg et al., 1987, for further discussion). In order to get rid of this noise McNamara and Zanetti (1988) proposed to use directly the mean occupation numbers instead of the Boolean fields. The lattice Boltzmann equation (LBE) had been applied already before by Wolfram (1986) and Frisch et al. (1987) as an analytical tool to calculate the viscosity coefficients of LGCA. McNamara and Zanetti for the first time used the LBE as a numerical scheme.

The mean occupation numbers F_i $(0 \leq F_i \leq 1)$ develop in time according to the following kinetic equation

$$F_i(r + c_i, t + 1) = F_i(r, t) + \Omega_i\left(\{F_j(r, t)\}\right), \qquad (5.1.6)$$

where the form of the collision operator Ω_i is identical to the arithmetic form of the microscopic collision operator of the corresponding LGCA. For FHP, for example, the Boolean (discrete) variables n_i in (3.2.17) are replaced by mean (continuous) occupation numbers F_i:

$$\begin{aligned}
\Omega_i(N) =\ & F_{i+1}F_{i+3}F_{i+5}(1 - F_i)(1 - F_{i+2})(1 - F_{i+4}) \\
& - F_iF_{i+2}F_{i+4}(1 - F_{i+1})(1 - F_{i+3})(1 - F_{i+5}) \\
& + \xi F_{i+1}F_{i+4}(1 - F_i)(1 - F_{i+2})(1 - F_{i+3})(1 - F_{i+5}) \quad (5.1.7) \\
& + (1 - \xi)F_{i+2}F_{i+5}(1 - F_i)(1 - F_{i+1})(1 - F_{i+3})(1 - F_{i+4}) \\
& - F_iF_{i+3}(1 - F_{i+1})(1 - F_{i+2})(1 - F_{i+4})(1 - F_{i+5}).
\end{aligned}$$

This type of LBM has been improved by Higuera and Jiménez (1989). They could show that the nonlinear collision operator which evaluation is time consuming can be approximated by a linear operator.

Further reading: Benzi et al. (1992) give an extensive review on LBMs based on the FHP and FCHC models.

5.2 BGK lattice Boltzmann model in 2D

Lattice Boltzmann models (LBMs) of the first generation are plagued by the same problems as the corresponding lattice-gas cellular automata except for the noise. Modern LBMs with Boltzmann distribution functions, several lattice speeds and BGK approximation of the collision operator are free of all problems mentioned in Table 5.1.1. In this chapter a detailed discussion of such a model in 2D will be given. The derivations of the equilibrium distributions and the Navier-Stokes equation will be presented in full length. We will refer to this model as the D2Q9-LBM.

A LBM has three main ingredients:

1. the lattice: D2Q9 (multi-speed model),

2. the equilibrium distributions: Maxwell (see below),

3. the kinetic equation: BGK approximation.

On the contrary, a LGCA is basically defined by the lattice and the collision rules.

Koelman (1991) defines his LBM for the Navier-Stokes equation over the lattice with the following lattice velocities c_i

$$
\begin{aligned}
c_0^K &= (0,0) \\
c_{1,2,3,4}^K &= (\pm a, \pm b) \\
c_{5,6}^K &= (\pm 2a, 0) \\
c_{7,8}^K &= (0, \pm 2b)
\end{aligned}
$$

where the lattice constants a and b are restricted by $a^2/3 \leq b^2 \leq 3a^2$. The special choice $a = b = \dfrac{1}{\sqrt{2}}c$ and rotation of the lattice velocities by $45°$ leads to the D2Q9 lattice (compare Section 3.3 and Fig. 5.2.1) with

$$
\begin{aligned}
c_0 &= (0,0) \\
c_{1,3}, c_{2,4} &= (\pm c, 0), (0, \pm c) \\
c_{5,6,7,8} &= (\pm c, \pm c).
\end{aligned}
\tag{5.2.1}
$$

Fig. 5.2.1. *The D2Q9 lattice.*

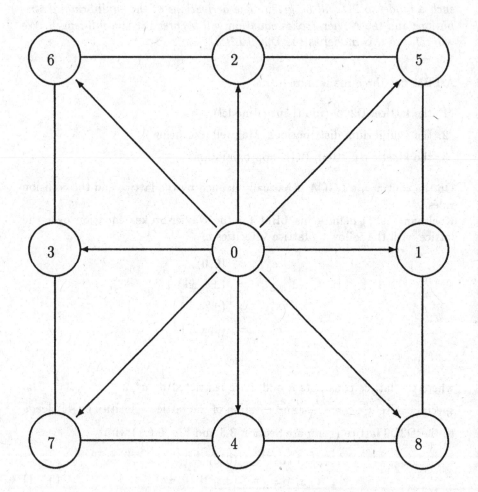

In what follows the D2Q9 lattice will be used exclusively.

The mass density, ρ, and the momentum density, j, are defined by sums over the distribution functions $F_i(x, t)$

$$\rho(x, t) = \sum_i F_i(x, t) \tag{5.2.2}$$

$$j(x, t) = \rho(x, t) v(x, t) = \sum_i c_i F_i(x, t) \tag{5.2.3}$$

For vanishing velocities a global equilibrium distribution W_i ("fluid at rest") is defined. In the vicinity (small Mach numbers) of this resting equilibrium, distribution functions can be written as sums of the W_i and small perturbations $f_i(x, t)$

$$F_i(x, t) = W_i + f_i(x, t) \tag{5.2.4}$$

with $|f_i(x, t)| << W_i$.

The W_i should be positive to assure positive mass density. They are chosen of Maxwell type in the following sense. The lattice velocity moments up to fourth order over the W_i shall be identical to the respective velocity moments over the Maxwell distribution

$$w_B(v) = \rho_0 \left(\frac{m}{2\pi k_B T} \right)^{D/2} \exp\left[-mv^2 / 2k_B T \right] \tag{5.2.5}$$

(D dimension, ρ_0 mass density, m particle mass, v particle speed, k_B Boltzmann constant, T temperature). Thus the odd moments vanish

$$\sum_i W_i c_{i\alpha} = 0$$

$$\sum_i W_i c_{i\alpha} c_{i\beta} c_{i\gamma} = 0$$

and the even moments read

$$\sum_i W_i = \int dv \, w_B(v) = \rho_0 \tag{5.2.6}$$

$$\sum_i W_i c_{i\alpha} c_{i\beta} = \int dv \, w_B(v) v_\alpha v_\beta = \rho_0 \frac{k_B T}{m} \delta_{\alpha\beta} \tag{5.2.7}$$

$$\sum_i W_i c_{i\alpha} c_{i\beta} c_{i\gamma} c_{i\delta} = \int dv \, w_B(v) v_\alpha v_\beta v_\gamma v_\delta$$

$$= \rho_0 \left(\frac{k_B T}{m} \right)^2 (\delta_{\alpha\beta} \delta_{\gamma\delta} + \delta_{\alpha\gamma} \delta_{\beta\delta} + \delta_{\alpha\delta} \delta_{\beta\gamma}) \tag{5.2.8}$$

Note that the constraint (5.2.8) is more rigorous than pure isotropy (compare Section 3.3).

Exercise 5.2.1. (**)
Calculate the velocity moments up to fourth order over the Maxwell distribution in $D \geq 2$ dimensions.

Nonnegative solutions of Eqs. (5.2.6 - 5.2.8) for the W_i can be found whenever the number of lattice velocities c_i is large enough. For the D2Q9 lattice (5.2.1) one obtains (see Section 5.2.1 for derivation):

$$W_0/\rho_0 = \frac{4}{9}$$

$$W_1/\rho_0 = \frac{1}{9}$$

$$W_2/\rho_0 = \frac{1}{36}$$

$$\frac{k_B T}{m} = \frac{c^2}{3}.$$

The evolution of the LBM consists of the recurring alternation between transition to the local equilibrium and propagation of the distributions to neighboring sites according to the lattice velocities. The BGK (compare Section 4.1.3) kinetic equation reads

$$F_i(x + c_i \Delta t, t + \Delta t) - F_i(x, t) = -\frac{\Delta t}{\tau} \left[F_i(x, t) - F_i^{(0)}(x, t) \right]$$

$$+ \frac{\Delta t c_{i\alpha}}{12c^2} \left[K_\alpha(x, t) + K_\alpha(x + c_i \Delta t, t + \Delta t) \right]$$

(5.2.9)

or

$$F_i(x + c_i \Delta t, t + \Delta t) = (1 - \omega) F_i(x, t) + \omega F_i^{(0)}(x, t)$$

$$+ \frac{\Delta t c_{i\alpha}}{12c^2} \left[K_\alpha(x, t) + K_\alpha(x + c_i \Delta t, t + \Delta t) \right]$$

(5.2.10)

where τ is the collision time, $\omega = \Delta t/\tau$ is the collision frequency, and K is an applied body force. The local equilibrium distributions $F_i^{(0)}$ depend only on the local values of mass and momentum density

$$F_i^{(0)}(x, t) = F_i^{(0)}(\rho(x, t), j(x, t)).$$

They can be derived by applying the maximum entropy principle under the constraints of mass and momentum conservation (see Section 5.2.2). Up to second order in j one obtains

$$F_i^{(0)}(\rho, j) = \frac{W_i}{\rho_0} \left\{ \rho + \frac{m}{k_B T} c_i \cdot j + \frac{m}{2\rho k_B T} \left[\frac{m}{k_B T} (c_i \cdot j)^2 - j^2 \right] \right\} \quad (5.2.11)$$

or more explicitly

$$F_i = \frac{4}{9}\rho \left[1 - \frac{3}{2} \frac{u^2}{c^2} \right] \qquad\qquad i = 0$$

$$F_i = \frac{1}{9}\rho \left[1 + 3\frac{c_i \cdot u}{c^2} + \frac{9}{2} \frac{(c_i \cdot u)^2}{c^4} - \frac{3}{2} \frac{u^2}{c^2} \right] \qquad i = 1,2,3,4$$

$$F_i = \frac{1}{36}\rho \left[1 + 3\frac{c_i \cdot u}{c^2} + \frac{9}{2} \frac{(c_i \cdot u)^2}{c^4} - \frac{3}{2} \frac{u^2}{c^2} \right] \qquad i = 5,6,7,8.$$

Application of the multi-scale technique (Chapman-Enskog expansion) yields the Navier-Stokes equation with pressure $p = \rho k_B T/m$, kinematic shear viscosity

$$\nu = \frac{k_B T}{m} \left(\frac{1}{\omega} - \frac{1}{2} \right) \Delta t = \frac{c^2}{3} \left(\frac{1}{\omega} - \frac{1}{2} \right) \Delta t = \frac{2 - \omega}{6\omega} c^2 \Delta t = \frac{c^2}{3} \left(\tau - \frac{\Delta t}{2} \right)$$
$$(5.2.12)$$

and an advection term with Galilean invariance[3]. The Galilean invariance breaking g-factor (compare Section 3.2) never arises (see Section 5.2.3).
The above given presentation of the LBM contains all informations necessary to set up the computer code. The algorithm proceeds as follows:

1. For given initial values of mass $\rho(x,t)$ and momentum density $j(x,t)$ calculate the equilibrium distributions $F_i^{(0)}(\rho(x,t), j(x,t))$ according to Eq. (5.2.11) and set $F_i = F_i^{(0)}$. Remark: Global equilibrium distributions (5.2.11) but with different values of ρ and j are used for local initialization, i.e. in the beginning there is a patchwork of local equilibria which is far from a global equilibrium.

2. Apply the kinetic equation (5.2.10), i.e. add the (non-equilibrium) distribution function $F_i(x,t)$ and the equilibrium distribution function $F_i^{(0)}(x,t)$ with the appropriate weights $(1-\omega)$ and ω, and then propagate it to the next neighbor (except for the distribution of 'rest particles' with $c_0 = 0$).

3. Calculate from the propagated distributions new values of $\rho(x,t)$ and $j(x,t)$ according to the definitions (5.2.2) und (5.2.3).

4. The next time step starts with the calculation of new equilibrium distributions. Thereafter proceed with the second step of the algorithm.

---

[3] But: There is always a disturbing fly or two in the ointment. Qian and Orszag (1993) and Qian and Zhou (1998) have shown that a higher order term (slightly compressible regime) leads to frame-velocity-dependent viscosity.

In BGK-LBMs collisions are not explicitly defined. They are kind of fictive and make themselves felt only by the transition to local equilibrium (the term $\omega \, F_i^{(0)}(\boldsymbol{x}, t)$ in the kinetic equation). On the other hand for LGCA the collisions are explicitly defined and the distribution functions are theoretical constructs which can not be observed on the lattice because of their continuous nature.

5.2.1 Derivation of the W_i

The D2Q9 model includes three different speeds (compare Table 5.2.1). The W_i for directions with identical speeds are equal for reason of symmetry.

Table 5.2.1. Lattice speeds, cells and W_i of the D2Q9 lattice.

c_i^2	cells	number of cells	W_i
0	0	1	W_0
1	$1, 2, 3, 4$	4	W_1
2	$5, 6, 7, 8$	4	W_2

The non-vanishing elements of the even moments up to fourth order read:

– 0. moment:

$$\sum_i W_i = W_0 + 4W_1 + 4W_2 = \rho_0 \qquad (5.2.13)$$

Remark: The only equation which includes W_0. It will be used to calculate W_0.

– 2. moment:

$$\sum_i c_{i1}^2 W_i = 2c^2 W_1 + 4c^2 W_2 = \rho_0 \frac{k_B T}{m} \qquad (5.2.14)$$

Remark: $(k_B T)/m$ will be calculated from Eqs. (5.2.14) and (5.2.15).

$$\sum_i c_{i2}^2 W_i = 2c^2 W_1 + 4c^2 W_2 = \rho_0 \frac{k_B T}{m}$$

Remark: This constraint is identical with (5.2.14).

– 4. Moment:

$$\sum_i c_{i1}^4 W_i = 2c^4 W_1 + 4c^4 W_2 = 3\rho_0 \left(\frac{k_B T}{m} \right)^2 \qquad (5.2.15)$$

Remark: W_1 will be calculated from Eq. (5.2.15).

$$\sum_i c_{i2}^4 W_i = 2c^4 W_1 + 4c^4 W_2 = 3\rho_0 \left(\frac{k_B T}{m}\right)^2$$

Remark: This constraint is identical with (5.2.15).

$$\sum_i c_{i1}^2 c_{i2}^2 W_i = 4c^4 W_2 = \rho_0 \left(\frac{k_B T}{m}\right)^2 \qquad (5.2.16)$$

Remark: W_2 will be calculated from Eq. (5.2.16).

The solution reads:

$$
\begin{aligned}
W_0/\rho_0 &= \frac{4}{9} \\
W_1/\rho_0 &= \frac{1}{9} \\
W_2/\rho_0 &= \frac{1}{36} \\
\frac{k_B T}{m} &= \frac{c^2}{3}.
\end{aligned}
$$

One readily confirms that the first and third moments of the lattice velocities over W_i vanish.

5.2.2 Entropy and equilibrium distributions

Koelman defines the *relative entropy*[4] density by

$$\boxed{S(\rho, \boldsymbol{j}) := -\frac{k}{m} \sum_i F_i^{(0)}(\rho, \boldsymbol{j}) \ln \frac{F_i^{(0)}(\rho, \boldsymbol{j})}{W_i}.} \qquad (5.2.17)$$

The weighting by $1/W_i$ in the logarithmic factor will lead to equilibrium distributions of the form $F_i^{(0)} = W_i e^{h(\rho, \boldsymbol{j}, \boldsymbol{c}_i)}$ and implies $S = 0$ for $F_i^{(0)} = W_i$. The equilibrium distributions $F_i^{(0)}$ will be calculated by maximizing the entropy for given constraints which for the case under consideration are the mass and momentum density

---

[4] Kullback (1959) and Cover and Thomas (1991) are standard references on relative entropy.

$$\rho(\rho, j) = \sum_i F_i^{(0)}(\rho, j)$$

$$j(\rho, j) = \sum_i c_i F_i^{(0)}(\rho, j).$$

The functional

$$\hat{S} := S + \tilde{A}\rho + \tilde{B} \cdot j$$

$$= -\frac{k}{m} \sum_i F_i^{(0)}(\rho, j) \ln \frac{F_i^{(0)}(\rho, j)}{W_i} + \tilde{A} \sum_i F_i^{(0)}(\rho, j) + \tilde{B} \sum_i c_i F_i^{(0)}(\rho, j)$$

encompasses the constraints coupled by Lagrange multipliers $\tilde{A}$ und $\tilde{B}$. The necessary conditions for an extremum of $\hat{S}$ read

$$\forall i: \quad \frac{\partial \hat{S}}{\partial F_i^{(0)}} = -\frac{k}{m}\left[\ln \frac{F_i^{(0)}}{W_i} + 1\right] + \tilde{A} + \tilde{B} \cdot c_i = 0. \tag{5.2.18}$$

The solutions of (5.2.18) are of the form

$$F_i^{(0)} = W_i e^{A(\rho,\, j) + B(\rho,\, j) \cdot c_i}$$

with

$$A = \frac{m}{k}\tilde{A} - 1, \quad \text{and} \quad B = \frac{m}{k}\tilde{B}.$$

A and B can be determined by Taylor series expansions of $F_i^{(0)}$ around $j = 0$ (same procedure as for the FHP lattice-gas cellular automata). Because of the symmetry of the D2Q9 lattice the ansatz

$$A(\rho, j) = A_0(\rho) + A_2(\rho)j^2 + \mathcal{O}(j^4)$$
$$B(\rho, j) = B_1(\rho)j + \mathcal{O}(j^3).$$

is sufficient. The expansion of $F_i^{(0)}$ around $j = 0$ reads

$$\frac{\partial F_i^{(0)}}{\partial j_\alpha} = (2A_2 j_\alpha + B_1 c_{i\alpha})F_i^{(0)}$$

$$\rightarrow B_1 c_{i\alpha} W_i e^{A_0} \quad \text{at } j = 0$$

$$\frac{\partial^2 F_i^{(0)}}{\partial j_\alpha^2} = \left[(2A_2 j_\alpha + B_1 c_{i\alpha})^2 + 2A_2\right] F_i^{(0)}$$

$$\rightarrow (B_1^2 c_{i\alpha}^2 + 2A_2)W_i e^{A_0} \quad \text{at } j = 0$$

$$\frac{\partial^2 F_i^{(0)}}{\partial j_\alpha \partial j_\beta} = (2A_2 j_\alpha + B_1 c_{i\alpha})(2A_2 j_\beta + B_1 c_{i\beta})F_i^{(0)}$$

$$\rightarrow B_1^2 c_{i\alpha} c_{i\beta} W_i e^{A_0} \quad \text{at } j = 0$$

and finally up to terms of second order in j

$$F_i^{(0)} = W_i e^{A_0} \left\{ 1 + B_1 c_i \cdot j + \frac{B_1^2}{2} (c_i \cdot j)^2 + A_2 j^2 \right\}. \qquad (5.2.19)$$

Now the definitions of ρ and j will be exploited to calculate the unknown coefficients $A_0(\rho)$, $A_2(\rho)$ and $B_1(\rho)$:

$$\sum_i F_i^{(0)} = \rho = e^{A_0} \left\{ \rho_0 + \frac{B_1^2}{2} \rho_0 \frac{k_B T}{m} j^2 + \rho_0 A_2 j^2 \right\} \qquad (5.2.20)$$

$$\sum_i c_i F_i^{(0)} = j = e^{A_0} B_1 \rho_0 \frac{k_B T}{m} j. \qquad (5.2.21)$$

The vector equation (5.2.21) reduces to a scalar constraint

$$B_1 = \frac{1}{e^{A_0}} \frac{m}{\rho_0 k_B T},$$

i.e. an auxiliary condition is required to solve for all three unknowns (A_0, A_2, B_1). Solving Eq. (5.2.20) for A_2 yields

$$A_2 = -\frac{B_1^2}{2} \frac{k_B T}{m} + \underbrace{\frac{1}{j^2} \left(\frac{\rho}{\rho_0 e^{A_0}} - 1 \right)}_{(*)}.$$

To obtain A_2 independent of j (as implied by the ansatz) the expression $(*)$ must vanish. This is the third constraint looked for. It immediately follows that

$$e^{A_0} = \frac{\rho}{\rho_0}, \quad B_1 = \frac{m}{\rho k_B T}, \quad \text{and} \quad A_2 = -\frac{1}{2\rho^2} \frac{m}{k_B T}.$$

Insertion into (5.2.19) finally yields the equilibrium distributions

$$F_i^{(0)}(\rho, j) = \frac{W_i}{\rho_0} \left\{ \rho + \frac{m}{k_B T} c_i \cdot j + \frac{m}{2\rho k_B T} \left[\frac{m}{k_B T} (c_i \cdot j)^2 - j^2 \right] \right\}. \qquad (5.2.22)$$

Koelman (1991) used $1/\rho_0$ instead of $1/\rho$ in the coefficient of the third term which is a good approximation for small Mach numbers.

Further reading: Karlin et al. (1998) construct local equilibrium functions, $F_i^{(0)}$, that maximize the entropy $S_K = \sum_i F_i^{(0)} \sqrt{F_i^{(0)}}$.

5.2.3 Derivation of the Navier-Stokes equations by multi-scale analysis

The derivation of the macroscopic equations (Navier-Stokes) proceeds in close analogy to the multi-scale analysis discussed in Section 4.2 and for the FHP model in Section 3.2. For the BGK-LBM the calculation of terms of second order in the expansion parameter ϵ is much simpler than for LGCA. Here the complete derivation of the Navier-Stokes equation will be given.

The distributions $F_i(x, t)$ are expanded around the equilibrium distributions $F_i^{(0)}(x, t)$

$$F_i(x, t) = F_i^{(0)}(x, t) + \epsilon F_i^{(1)}(x, t) + \epsilon^2 F_i^{(2)} + \mathcal{O}(\epsilon^3) \tag{5.2.23}$$

with

$$\sum_i F_i^{(1)}(x, t) = 0, \quad \sum_i c_i F_i^{(1)}(x, t) = 0,$$

$$\tag{5.2.24}$$

$$\sum_i F_i^{(2)}(x, t) = 0, \quad \sum_i c_i F_i^{(2)}(x, t) = 0,$$

i.e. the perturbations $F_i^{(1)}(x, t)$ and $F_i^{(2)}(x, t)$ do not contribute to the mass and momentum density. The small expansion parameter ϵ can be viewed as the Knudsen number K_n which is the ratio between the mean free path and the characteristic length scale of the flow.

The left hand side of the kinetic equation (5.2.10) and the forcing term are expanded into a Taylor series up to terms of second order ($\Delta x_i = \Delta t \cdot c_i$):

$$F_i(x + c_i \Delta t, t + \Delta t) = F_i(x, t) + \Delta t \partial_t F_i + \Delta t c_{i\alpha} \partial_{x_\alpha} F_i$$

$$\tag{5.2.25}$$

$$+ \frac{(\Delta t)^2}{2} \left[\partial_t \partial_t F_i + 2 c_{i\alpha} \partial_t \partial_{x_\alpha} F_i + c_{i\alpha} c_{i\beta} \partial_{x_\alpha} \partial_{x_\beta} F_i \right] + \mathcal{O}\left(\partial^3 F_i\right)$$

and

$$\frac{\Delta t}{c^2} \frac{c_{i\alpha}}{12} \left[K_\alpha(x, t) + K_\alpha(x + c_i \Delta t, t + \Delta t) \right]$$

$$\tag{5.2.26}$$

$$= \frac{\Delta t}{c^2} \frac{c_{i\gamma}}{6} K_\gamma(x, t) + \frac{(\Delta t)^2}{c^2} \frac{c_{i\gamma}}{12} \left[\partial_t K_\gamma + c_{i\beta} \partial_{x_\beta} K_\gamma \right] + \mathcal{O}\left[(\Delta t)^3\right].$$

As for the FHP model two time scales and one spatial scale with the following scaling will be introduced

$$\partial_t \quad \rightarrow \quad \epsilon \partial_t^{(1)} + \epsilon^2 \partial_t^{(2)}$$

$$\partial_{x_\alpha} \quad \rightarrow \quad \epsilon \partial_{x_\alpha}^{(1)}.$$

$(5.2.27)$

Conservation of mass and momentum. The expansions given above are substituted into the kinetic equation (Eq. 5.2.10)

$$
\begin{aligned}
0 = \quad & F_i(\boldsymbol{x} + \boldsymbol{c}_i \Delta t, t + \Delta t) - F_i(\boldsymbol{x}, t) + \omega \left[F_i(\boldsymbol{x}, t) - F_i^{(0)}(\boldsymbol{x}, t) \right] \\
& - \frac{\Delta t}{c^2} \frac{c_{i\alpha}}{12} \left[K_\alpha(\boldsymbol{x}, t) + K_\alpha(\boldsymbol{x} + \boldsymbol{c}_i \Delta t, t + \Delta t) \right]
\end{aligned}
$$

which leads to

$$
\begin{aligned}
0 \underbrace{=}_{(5.2.25)} \quad & F_i(\boldsymbol{x}, t) + \Delta t \partial_t F_i + \Delta t c_{i\gamma} \partial_{x_\gamma} F_i \\
& + \frac{(\Delta t)^2}{2} \left[\partial_t \partial_t F_i + 2 c_{i\gamma} \partial_t \partial_{x_\gamma} F_i + c_{i\beta} c_{i\gamma} \partial_{x_\beta} \partial_{x_\gamma} F_i \right] \\
& + \mathcal{O} \left[\partial^3 F_i \right] - F_i(\boldsymbol{x}, t) + \omega \left[F_i(\boldsymbol{x}, t) - F_i^{(0)}(\boldsymbol{x}, t) \right] \\
& - \frac{\Delta t}{c^2} \frac{c_{i\gamma}}{6} K_\gamma(\boldsymbol{x}, t) - \frac{(\Delta t)^2}{c^2} \frac{c_{i\gamma}}{12} \left[\partial_t K_\gamma + c_{i\beta} \partial_{x_\beta} K_\gamma \right] \\
& + \mathcal{O} \left[(\Delta t)^3 \right] \\
\underbrace{=}_{(5.2.23),(5.2.27)} \quad & \epsilon \Delta t \left[\partial_t^{(1)} F_i^{(0)} + c_{i\gamma} \partial_{x_\gamma}^{(1)} F_i^{(0)} \right] \\
& + \epsilon^2 \Delta t \left[\partial_t^{(1)} F_i^{(1)} + \partial_t^{(2)} F_i^{(0)} + c_{i\gamma} \partial_{x_\gamma}^{(1)} F_i^{(1)} \right] \\
& + \epsilon^2 \frac{(\Delta t)^2}{2} \left[\partial_t^{(1)} \partial_t^{(1)} F_i^{(0)} + 2 c_{i\gamma} \partial_t^{(1)} \partial_{x_\gamma}^{(1)} F_i^{(0)} \right. \\
& \left. + c_{i\beta} c_{i\gamma} \partial_{x_\beta}^{(1)} \partial_{x_\gamma}^{(1)} F_i^{(0)} \right] + \epsilon \omega F_i^{(1)} \\
& + \epsilon^2 \omega F_i^{(2)} - \epsilon \frac{\Delta t}{c^2} \frac{c_{i\gamma}}{6} K_\gamma \\
& - \epsilon^2 \frac{(\Delta t)^2}{c^2} \frac{c_{i\gamma}}{12} \left[\partial_t^{(1)} K_\gamma + c_{i\beta} \partial_{x_\beta}^{(1)} K_\gamma \right] + \mathcal{O} \left[\epsilon^3 \right]
\end{aligned}
$$

Please note that the leading order of the body forcing was set proportional to ϵ. Sorting according to orders in ϵ yields

$$0 = \epsilon E_i^{(0)} + \epsilon^2 E_i^{(1)} + \mathcal{O}\left[\epsilon^3\right] \tag{5.2.28}$$

with

$$
\begin{aligned}
E_i^{(0)} &= \partial_t^{(1)} F_i^{(0)} + c_{i\gamma}\partial_{x_\gamma}^{(1)} F_i^{(0)} + \frac{\omega}{\Delta t} F_i^{(1)} - \frac{c_{i\gamma}}{6c^2} K_\gamma \tag{5.2.29} \\[2mm]
E_i^{(1)} &= \partial_t^{(1)} F_i^{(1)} + \partial_t^{(2)} F_i^{(0)} + c_{i\gamma}\partial_{x_\gamma}^{(1)} F_i^{(1)} + \frac{\Delta t}{2}\partial_t^{(1)}\partial_t^{(1)} F_i^{(0)} \\[2mm]
&\quad + \Delta t c_{i\gamma}\partial_t^{(1)}\partial_{x_\gamma}^{(1)} F_i^{(0)} + \frac{\Delta t}{2} c_{i\beta}c_{i\gamma}\partial_{x_\beta}^{(1)}\partial_{x_\gamma}^{(1)} F_i^{(0)} \tag{5.2.30} \\[2mm]
&\quad + \frac{\omega}{\Delta t} F_i^{(2)} - \Delta t \frac{c_{i\gamma}}{12c^2}\partial_t^{(1)} K_\gamma - \Delta t \frac{c_{i\gamma}}{12c^2} c_{i\beta}\partial_{x_\beta}^{(1)} K_\gamma .
\end{aligned}
$$

We will now calculate the zeroth and first lattice velocity moments of $E_i^{(0)}$ and $E_i^{(1)}$ (conservation of mass and momentum density).

Terms of first order in ϵ. The zeroth and first lattice velocity moments of $E_i^{(0)}$ can be readily calculated:

$$
\begin{aligned}
\sum_i E_i^{(0)} &= \sum_i \left\{ \partial_t^{(1)} F_i^{(0)} + c_{i\gamma}\partial_{x_\gamma}^{(1)} F_i^{(0)} + \frac{\omega}{\Delta t} F_i^{(1)} - \frac{c_{i\gamma}}{6c^2} K_\gamma \right\} \\[2mm]
&= \partial_t^{(1)}\rho + \partial_{x_\gamma}^{(1)} j_\gamma
\end{aligned}
$$

$$
\begin{aligned}
\sum_i c_{i\alpha} E_i^{(0)} &= \sum_i c_{i\alpha} \left\{ \partial_t^{(1)} F_i^{(0)} + c_{i\beta}\partial_{x_\beta}^{(1)} F_i^{(0)} + \frac{\omega}{\Delta t} F_i^{(1)} - \frac{c_{i\beta}}{6c^2} K_\beta \right\} \\[2mm]
&= \partial_t^{(1)} j_\alpha + \partial_{x_\beta}^{(1)} P_{\alpha\beta}^{(0)} - K_\alpha
\end{aligned}
\tag{5.2.31}
$$

where we have used

$$\sum_i c_{i\alpha} c_{i\beta} = 6c^2 \delta_{\alpha\beta}. \tag{5.2.32}$$

The momentum flux tensor in first order of ϵ reads

$$P_{\alpha\beta}^{(0)} := \sum_i c_{i\alpha} c_{i\beta} F_i^{(0)} = \frac{1}{\rho}\begin{pmatrix} j_1^2 & j_1 j_2 \\ j_1 j_2 & j_2^2 \end{pmatrix} + p\delta_{\alpha\beta} \tag{5.2.33}$$

where $p = \dfrac{k_B T}{m}\rho$ is the pressure (a detailed calculation of $P_{\alpha\beta}^{(0)}$ will be given below).

Thus to first order in ϵ we obtain the *continuity equation*

$$\boxed{\partial_t \rho + \boldsymbol{\nabla} \cdot \boldsymbol{j} = 0} \tag{5.2.34}$$

and (in the incompressible limit: $\rho =$ constant) the *Euler equation* (Navier-Stokes equation without friction):

$$\boxed{\frac{\partial \boldsymbol{u}}{\partial t} + \boldsymbol{u}\boldsymbol{\nabla}\boldsymbol{u} = -\frac{1}{\rho}\boldsymbol{\nabla}p + \boldsymbol{K}.} \tag{5.2.35}$$

Calculation of the momentum flux tensor. In the calculation of $P_{\alpha\beta}^{(0)}$ the detailed form of the equilibrium distribution has to be taken into account for the first time. Four different summands $P_{\alpha\beta}^{(0.1)}$ to $P_{\alpha\beta}^{(0.4)}$ have to be evaluated. The first term reads:

$$P_{\alpha\beta}^{(0.1)} := \frac{\rho}{\rho_0}\sum_i c_{i\alpha}c_{i\beta}W_i = \rho\underbrace{\frac{k_B T}{m}}_{=\,p}\delta_{\alpha\beta} = p\delta_{\alpha\beta}.$$

For the pressure $p = \dfrac{k_B T}{m}\rho = \dfrac{c^2}{3}\rho$ the speed of sound c_s is given by

$$c_s = \sqrt{\frac{dp}{d\rho}} = \frac{c}{\sqrt{3}}.$$

The second summand

$$P_{\alpha\beta}^{(0.2)} := \frac{m}{\rho_0 k_B T}\sum_i c_{i\alpha}c_{i\beta}(\boldsymbol{c}_i \cdot \boldsymbol{j})W_i = 0$$

vanishes because it is an odd moment in $\boldsymbol{c}_i$ over W_i. The third summand reads:

$$P_{\alpha\beta}^{(0.3)} := \frac{1}{2\rho_0\rho}\left(\frac{m}{k_B T}\right)^2 \sum_i c_{i\alpha}c_{i\beta}\left(\boldsymbol{c}_i \cdot \boldsymbol{j}\right)^2 W_i.$$

Prefactor:

$$\frac{1}{2\rho_0\rho}\left(\frac{m}{k_B T}\right)^2 \cdot \rho_0\left(\frac{k_B T}{m}\right)^2 = \frac{1}{2\rho}$$

Tensor of fourth rank:

$$T_{\alpha\beta\gamma\delta} = \delta_{\alpha\beta}\delta_{\gamma\delta} + \delta_{\alpha\gamma}\delta_{\beta\delta} + \delta_{\alpha\delta}\delta_{\beta\gamma} \tag{5.2.36}$$

$-\ \alpha = \beta = 1:$

  $*\ \gamma = \delta = 1 \rightarrow T_{1111} = 3 \rightarrow \dfrac{1}{2\rho}3j_1^2$

  $*\ \gamma = \delta = 2 \rightarrow T_{1122} = 1 \rightarrow \dfrac{1}{2\rho}j_2^2$

  $*\ \gamma \ne \delta \rightarrow T_{1112} = T_{1121} = 0$

$-\ \alpha = \beta = 2: \rightarrow \dfrac{1}{2\rho}\left(j_1^2 + 3j_2^2\right)$

$-\ \alpha = 1\ \beta = 2:$

  $-\ \gamma = \delta \rightarrow T_{1211} = T_{1222} = 0$

  $-\ \gamma \ne \delta \rightarrow T_{1212} = T_{1221} = 1 \rightarrow \dfrac{1}{2\rho} \cdot 2j_1 j_2$

$$P_{\alpha\beta}^{(0.3)} = \frac{1}{2\rho}j^2 \delta_{\alpha\beta} + \frac{1}{\rho}\begin{pmatrix} j_1^2 & j_1 j_2 \\ j_1 j_2 & j_2^2 \end{pmatrix}$$

The fourth summand reads:

$$
\begin{aligned}
P_{\alpha\beta}^{(0.4)} &:= -\frac{1}{2\rho_0\rho}\frac{m}{k_B T}j^2 \sum_i c_{i\alpha}c_{i\beta}W_i \\
&= -\frac{1}{2\rho}j^2 \delta_{\alpha\beta}
\end{aligned}
$$

Summing up all terms yields

$$P_{\alpha\beta}^{(0)} = \frac{1}{\rho}\begin{pmatrix} j_1^2 & j_1 j_2 \\ j_1 j_2 & j_2^2 \end{pmatrix} + p\delta_{\alpha\beta} \qquad (5.2.37)$$

The pressure of the D2Q9-LBM does not depend explicitly on the flow speed (see to the contrary the FHP model) because the summand $P_{\alpha\beta}^{(0.4)}$ is exactly compensated by a part of $P_{\alpha\beta}^{(0.3)}$.

Terms of second order in ϵ: mass. As for lattice-gas cellular automata one has to take into account terms of second order in ϵ to derive the dissipative terms of the macroscopic equations. Mass conservation leads to

$$0 = \sum_i E_i^{(1)}$$

$$= \sum_i \partial_t^{(1)} F_i^{(1)} + \partial_t^{(2)} F_i^{(0)} + c_{i\gamma} \partial_{x_\gamma}^{(1)} F_i^{(1)} + \frac{\Delta t}{2} \partial_t^{(1)} \partial_t^{(1)} F_i^{(0)}$$

$$+ \Delta t c_{i\gamma} \partial_t^{(1)} \partial_{x_\gamma}^{(1)} F_i^{(0)} + \frac{\Delta t}{2} c_{i\beta} c_{i\gamma} \partial_{x_\beta}^{(1)} \partial_{x_\gamma}^{(1)} F_i^{(0)} \qquad (5.2.38)$$

$$+ \frac{\omega}{\Delta t} F_i^{(2)} - \Delta t \frac{c_{i\gamma}}{12 c^2} \partial_t^{(1)} K_\gamma - \Delta t \frac{c_{i\gamma}}{12 c^2} c_{i\beta} \partial_{x_\beta}^{(1)} K_\gamma.$$

The following summands vanish:

$$\partial_t^{(1)} \sum_i F_i^{(1)} \underset{(5.2.24)}{=} 0,$$

$$\partial_{x_\gamma}^{(1)} \sum_i c_{i\gamma} F_i^{(1)} \underset{(5.2.24)}{=} 0,$$

$$\frac{\omega}{\Delta t} \sum_i F_i^{(2)} \underset{(5.2.24)}{=} 0,$$

and

$$\frac{\Delta t}{12 c^2} \partial_t^{(1)} K_\gamma \sum_i c_{i\gamma} = 0.$$

The second term of Eq. (5.2.38) is the time derivative of the density:

$$\partial_t^{(2)} \sum_i F_i^{(0)} = \partial_t^{(2)} \rho. \qquad (5.2.39)$$

The spatial gradient of the body force yields

$$-\frac{\Delta t}{12 c^2} \partial_{x_\alpha}^{(1)} K_\beta \sum_i c_{i\alpha} c_{i\beta} = -\frac{\Delta t}{2} \partial_{x_\alpha}^{(1)} K_\alpha. \qquad (5.2.40)$$

In the transformation of the following summands the time derivatives are substituted by spatial derivative using (5.2.34) and (5.2.31):

$$\frac{\Delta t}{2} \partial_t^{(1)} \partial_t^{(1)} \sum_i F_i^{(0)} = \frac{\Delta t}{2} \partial_t^{(1)} \partial_t^{(1)} \rho = -\frac{\Delta t}{2} \partial_t^{(1)} \partial_{x_\beta}^{(1)} \rho u_\beta$$

$$= \frac{\Delta t}{2} \partial_{x_\alpha}^{(1)} \partial_{x_\beta}^{(1)} P_{\alpha\beta}^{(0)} - \frac{\Delta t}{2} \partial_{x_\alpha}^{(1)} K_\alpha$$

$$\Delta t \partial_t^{(1)} \partial_{x_\alpha}^{(1)} \sum_i c_{i\alpha} F_i^{(0)} = \Delta t \partial_{x_\alpha}^{(1)} \partial_t^{(1)} (\rho u_\alpha)$$

$$= -\Delta t \partial_{x_\alpha}^{(1)} \partial_{x_\beta}^{(1)} P_{\alpha\beta}^{(0)} + \Delta t \partial_{x_\alpha}^{(1)} K_\alpha$$

$$\frac{\Delta t}{2} \partial_{x_\alpha}^{(1)} \partial_{x_\beta}^{(1)} \sum_i c_{i\alpha} c_{i\beta} F_i^{(0)} = \frac{\Delta t}{2} \partial_{x_\alpha}^{(1)} \partial_{x_\beta}^{(1)} P_{\alpha\beta}^{(0)}$$

The sum of these three terms yields $\Delta t \partial_{x_\alpha}^{(1)} K_\alpha / 2$ which cancels with the term derived from the body force gradient (Eq. 5.2.40), i.e. inclusion of the spatial gradient of the body force is essential to ensure that there is no mass diffusion:

$$\boxed{\partial_t^{(2)} \rho = 0.}$$ (5.2.41)

Terms of second order in ϵ: momentum. Conservation of momentum leads to

$$
\begin{aligned}
0 &= \sum_i c_{i\alpha} E_i^{(1)} \\
&= \sum_i c_{i\alpha} \partial_t^{(1)} F_i^{(1)} + c_{i\alpha} \partial_t^{(2)} F_i^{(0)} + c_{i\alpha} c_{i\beta} \partial_{x_\beta}^{(1)} F_i^{(1)} + \frac{\Delta t}{2} c_{i\alpha} \partial_t^{(1)} \partial_t^{(1)} F_i^{(0)} \\
&\quad + \Delta t c_{i\alpha} c_{i\gamma} \partial_t^{(1)} \partial_{x_\gamma}^{(1)} F_i^{(0)} + \frac{\Delta t}{2} c_{i\alpha} c_{i\beta} c_{i\gamma} \partial_{x_\beta}^{(1)} \partial_{x_\gamma}^{(1)} F_i^{(0)} \\
&\quad + \frac{\omega}{\Delta t} c_{i\alpha} F_i^{(2)} - \Delta t c_{i\alpha} \frac{c_{i\gamma}}{12c^2} \partial_t^{(1)} K_\gamma - c_{i\alpha} c_{i\beta} c_{i\gamma} \frac{\Delta t}{12c^2} \partial_{x_\beta}^{(1)} K_\gamma.
\end{aligned}
$$

An approximation of $F_i^{(1)}$ can be derived from the first order in ϵ: $E_i^{(0)} = 0$ $\Rightarrow$

$$\boxed{F_i^{(1)}(\boldsymbol{x}, t) = -\frac{\Delta t}{\omega} \partial_t^{(1)} F_i^{(0)} - \frac{\Delta t}{\omega} c_{i\gamma} \partial_{x_\gamma}^{(1)} F_i^{(0)} + \frac{\Delta t\, c_{i\gamma}}{6c^2 \omega} K_\gamma,}$$ (5.2.42)

i.e. deviations from local equilibrium are driven by gradients in time and space (compare Eq. 4.2.25) and by applied body forces.

In the transformation of the following summands terms of the order $\mathcal{O}(j^2)$ will be neglected (indicated by $\approx$ instead of $=$). Thus the Navier-Stokes equation will be recovered in the limit of *low Mach numbers* only.

$$\partial_t^{(1)} \sum_i c_{i\alpha} F_i^{(1)} \underset{(5.2.24)}{\approx} 0$$

$$\partial_t^{(2)} \sum_i c_{i\alpha} F_i^{(0)} = \partial_t^{(2)} j_\alpha$$ (5.2.43)

$$\partial_{x_\beta}^{(1)} \sum_i c_{i\alpha} c_{i\beta} F_i^{(1)} \underset{(5.2.42)}{\approx} \underbrace{-\frac{\Delta t}{\omega} \partial_t^{(1)} \partial_{x_\beta}^{(1)} \sum_i c_{i\alpha} c_{i\beta} F_i^{(0)}}_{(*)}$$

$$\underbrace{-\frac{\Delta t}{\omega} \partial_{x_\beta}^{(1)} \partial_{x_\gamma}^{(1)} \sum_i c_{i\alpha} c_{i\beta} c_{i\gamma} F_i^{(0)}}_{(**)}$$

$$+\frac{\Delta t}{6c^2\omega}\partial_{x_\beta}^{(1)}K_\gamma\underbrace{\sum_i c_{i\alpha}c_{i\beta}c_{i\gamma}}_{=0}$$

$$
\begin{aligned}
\frac{\Delta t}{2}\partial_t^{(1)}\partial_t^{(1)}\sum_i c_{i\alpha}F_i^{(0)} &= \frac{\Delta t}{2}\partial_t^{(1)}\partial_t^{(1)}j_\alpha \\
&= -\frac{\Delta t}{2}\partial_t^{(1)}\partial_{x_\beta}^{(1)}P_{\alpha\beta}^{(0)} + \frac{\Delta t}{2}\partial_t^{(1)}K_\alpha \\
&\approx -\frac{\Delta t}{2}\frac{k_BT}{m}\partial_t^{(1)}\partial_{x_\beta}^{(1)}\rho\delta_{\alpha\beta} + \frac{\Delta t}{2}\partial_t^{(1)}K_\alpha \\
&\Rightarrow \frac{\Delta t}{2}\frac{k_BT}{m}\boldsymbol{\nabla}\boldsymbol{\nabla}\cdot\boldsymbol{j} + \frac{\Delta t}{2}\partial_t\boldsymbol{K} \qquad (5.2.44)
\end{aligned}
$$

$$\underbrace{\frac{\Delta t}{2}\partial_{x_\beta}^{(1)}\partial_{x_\gamma}^{(1)}\sum_i c_{i\alpha}c_{i\beta}c_{i\gamma}F_i^{(0)}}_{(**)}$$

$$\underbrace{\Delta t\partial_{x_\beta}^{(1)}\partial_t^{(1)}\sum_i c_{i\alpha}c_{i\beta}F_i^{(0)}}_{(*)}$$

$$\frac{\omega}{\Delta t}\sum_i c_{i\alpha}F_i^{(2)}\underbrace{=}_{(5.2.24)}0$$

The term derived from the time derivative of the body force

$$-\frac{\Delta t}{12c^2}\partial_t^{(1)}K_\beta\sum_i c_{i\alpha}c_{i\beta} = -\frac{\Delta t}{2}\partial_t^{(1)}K_\alpha \Rightarrow -\frac{\Delta t}{2}\partial_t\boldsymbol{K} \qquad (5.2.45)$$

cancels with the term in Eq. (5.2.44), i.e. inclusion of the time derivative of the body force is essential to obtain the exact form of the Navier-Stokes equation.

$$-\frac{\Delta t}{12c^2}\partial_{x_\beta}^{(1)}K_\gamma\sum_i c_{i\alpha}c_{i\beta}c_{i\gamma} = 0$$

Summation of the (*)-terms results in

$$\Delta t\left(1-\frac{1}{\omega}\right)\partial_{x_\beta}^{(1)}\partial_t^{(1)}\underbrace{\sum_i c_{i\alpha}c_{i\beta}F_i^{(0)}}_{\approx\frac{k_BT}{m}\rho\delta_{\alpha\beta}} \Rightarrow -\Delta t\left(1-\frac{1}{\omega}\right)\frac{k_BT}{m}\boldsymbol{\nabla}\boldsymbol{\nabla}\cdot\boldsymbol{j}. \qquad (5.2.46)$$

The (**)-terms lead to

$$\Delta t \quad \left(\frac{1}{2} - \frac{1}{\omega}\right) \partial_{x_\beta}^{(1)} \partial_{x_\gamma}^{(1)} \sum_i c_{i\alpha} c_{i\beta} c_{i\gamma} F_i^{(0)}$$

$$= \quad \Delta t \left(\frac{1}{2} - \frac{1}{\omega}\right) \frac{m}{k_B T} \frac{1}{\rho_0} \partial_{x_\beta}^{(1)} \partial_{x_\gamma}^{(1)} \sum_i W_i c_{i\alpha} c_{i\beta} c_{i\gamma} (\boldsymbol{c}_i \cdot \boldsymbol{j})$$

$$\Rightarrow \quad \Delta t \left(\frac{1}{2} - \frac{1}{\omega}\right) \frac{k_B T}{m} \left(\nabla^2 \boldsymbol{j} + 2 \boldsymbol{\nabla}\boldsymbol{\nabla} \cdot \boldsymbol{j}\right). \tag{5.2.47}$$

By adding up (5.2.43 – 5.2.47) one finally obtains

$$\boxed{\partial_t^{(2)} \boldsymbol{j} = \Delta t \left(\frac{1}{\omega} - \frac{1}{2}\right) \frac{k_B T}{m} \left(\nabla^2 \boldsymbol{j} + \boldsymbol{\nabla}\boldsymbol{\nabla} \cdot \boldsymbol{j}\right),} \tag{5.2.48}$$

i.e. dynamic shear and compression viscosity are equal:

$$\mu_S = \mu_K = \Delta t \left(\frac{1}{\omega} - \frac{1}{2}\right) \frac{k_B T}{m}. \tag{5.2.49}$$

The viscosities can be tuned by an appropriate choice of the collision parameter ω. The equation for iterative solution of linear systems by the simultaneous over-relaxation (SOR) is of the same form as the BGK kinetic equation. The SOR method is convergent for $0 < \omega < 2$ (Kahan, 1958). This range of ω corresponds to positive viscosities in the BGK-LBM.

The sum of the first and second order terms yields (in the incompressible limit) the Navier-Stokes equation:

$$\boxed{\boldsymbol{\nabla} \cdot \boldsymbol{u} = 0} \tag{5.2.50}$$

and

$$\boxed{\partial_t \boldsymbol{u} + (\boldsymbol{u}\boldsymbol{\nabla}) \boldsymbol{u} = -\frac{1}{\rho} \boldsymbol{\nabla} p + \nu \nabla^2 \boldsymbol{u} + \boldsymbol{K}} \tag{5.2.51}$$

with the kinematic shear viscosity

$$\nu = \Delta t \left(\frac{1}{\omega} - \frac{1}{2}\right) \frac{c^2}{3} = \frac{c^2}{3} \left(\tau - \frac{\Delta t}{2}\right). \tag{5.2.52}$$

5.2.4 Storage demand

For the special case $\omega = 1$ the kinetic equation reduces to

$$F_i(\boldsymbol{x} + \boldsymbol{c}_i \Delta t, t + \Delta t) = F_i^{(0)}(\boldsymbol{x}, t)$$

and the storage demand for ρ, $\boldsymbol{j}$, 9 distributions $F_i^{(eq)}$ and their propagated values F_i, i.e. 21 2D arrays, seems to be quite large. The required memory can, however, be drastically reduced by the following two procedures:

1. One after the next $F_i^{(eq)}$ can be evaluated, propagated and will contribute (independent of the other F_j) to the values of ρ and j at the next time level. Only 7 arrays, namely ρ and j at two time levels and one array F_i, are required, i.e. the reduction factor is $21/7 = 3$.

2. Because the collisions are strictly local and the propagation is almost local (to next neighbor nodes) the domain can be partitioned into subdomains which can be updated one after the other using smaller auxiliary arrays.

Procedure 1 leads to some increase in the length of the code but seems to be worth the effort especially in 3D where the reduction factor is even larger. The second procedure requires some additional coding to treat the boundaries of the subdomains but offers a further reduction in memory by a factor of almost 2.

The storage demand for BGK-LBM with $\omega \neq 1$ is substantially larger because one has to keep all (non-equilibrium) distributions F_i. The adventage is the tunable viscosity which allows simulation of flows with high Reynolds numbers.

5.2.5 Simulation of two-dimensional decaying turbulence

In 1994 Martinez et al. published a milestone paper. They simulated two-dimensional decaying turbulence over a square with periodic boundary conditions by the BGK lattice Boltzmann model over the D2Q9 lattice and compared this method to a spectral model[5] which is very efficient over this simple domain. The initial velocity distribution consists of two shear layers (realized by a truncated spectral representation of δ functions at $y = \pi/2$ and $y = 3\pi/2$) plus some noise over the whole domain. The characteristic velocity U is defined as the root mean square value

$$U = \sqrt{\langle u^2 \rangle} = 0.04 \qquad (5.2.53)$$

und the characteristic length L is given by

$$L = \frac{512}{2\pi} \qquad (5.2.54)$$

where 512 is the number of grid points in each dimension. The viscosity parameter ω has been chosen such that the Reynolds number

$$R_e = \frac{UL}{\nu(\omega)} \qquad (5.2.55)$$

---

[5] The spectral model actually does not solve the Navier-Stokes equation but the vorticity equation $\dfrac{\partial \varpi}{\partial t} + v \cdot \nabla \varpi = \nu \nabla^2 \varpi$ where $\varpi = (\nabla \times v)_z$ is the z-component of the relative vorticity.

is 10000.

The following Figures (5.2.2 - 5.2.5) show isocontours of the vorticity $a = (\nabla \times u)_z$ at four dimensionless times. Dashed lines indicate negative values. The upper plots show the results of the spectral model and the lower plots those of the lattice Boltzmann model. The features compare very well which gives confidence in both methods.

What is most surprising: the LBM is as fast as the spectral model! And the LBM keeps its efficiency in more complex geometries (porous media, for example) whereas spectral models may not be applicable anymore. These simulations clearly demonstrated the great potential of BGK-LBMs.

Fig. 5.2.2. *Isovorticity contour plots for time 1. Dashed lines correspond to negative values of vorticity. The values for the contours are the same for all cases (Figures 5.2.2 – 5.2.5). Strikingly similar features can be found for the lattice Boltzmann simulation, as compared with the spectral simulation. (SP refers to the spectral sumulation and LBE to the lattice Boltzmann simulation; Martinez et al., 1994)*

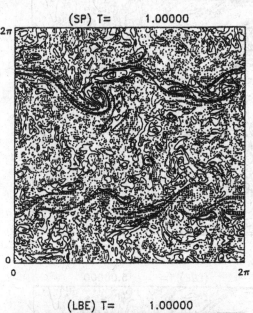

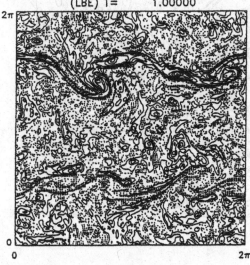

Fig. 5.2.3. *Same as in Fig. 5.2.2 for time 5 (Martinez et al., 1994).*

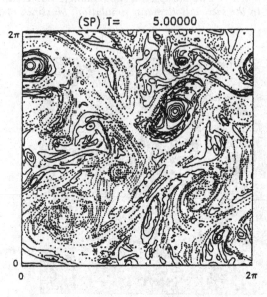

Fig. 5.2.4. *Same as in Fig. 5.2.2 for time 17 (Martinez et al., 1994).*

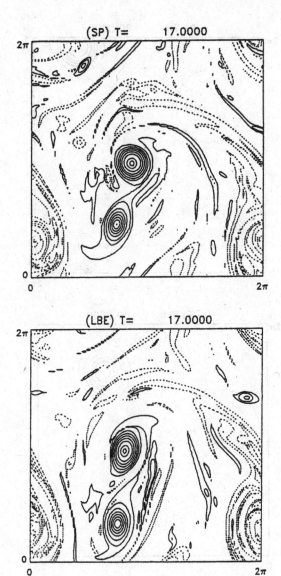

Fig. 5.2.5. *Same as in Fig. 5.2.2 for time 80 (Martinez et al., 1994).*

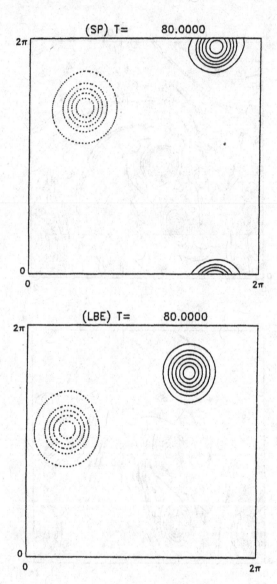

5.2.6 Boundary conditions for LBM

> "To a certain degree, achieving self-consistent boundary conditions
> with a given accuracy is as important as developing numerical
> schemes themselves."
> Chen et al. (1996)

The implementation of boundary conditions for LGCA has been discussed
in Section 3.2.7. The same princles as outlined there can be applied also to
LBM. In the following a self-contained discussion of no-slip and slip boundary
conditions shall be given even risking the repetition of parts of Section 3.2.7.
LGCA and LBM seem to be very attractive because the apparent ease to
implement boundary conditions even in complicated geometry like porous
media. You often can find quotes like 'difficult geometrical boundary conditions are readily handled' in the introduction of an article only to find
out that the authors performed some simulation over a square with periodic
boundary conditions. Much more realistic is the remark of He et al. (1997)
that "the real hydrodynamic boundary conditions have not been fully understood". Accordingly various implementations of boundary conditions will
be discussed here. Some of them have to be ruled out because they are only
of first order of accuracy. You have to find out which of the remaining ones
gives satisfying results for the actual flow problem considered.

In general, there are two ways to define a boundary: the boundary curve may
include grid nodes ('*node boundary*'; the nodes on the boundary are called
'*boundary nodes*'; Skordos, 1993; Inamuro et al., 1995; Noble et al., 1995a,b,
1996) or passes through the midpoints of links between nodes ('*link boundary*'; Cornubert et al., 1991; Ginzbourg and Adler, 1994; Ladd, 1994a,b).
Node boundaries are appropriate for periodic and inflow boundary conditions.

No-slip boundary condition. Inamuro et al. (1995) He et al. (1997)

1. The *complete bounceback scheme*: Instead of collision assign each F_i the
   value of the F_i of its opposite direction, i.e. for a point on the lower
   boundary:

$$\text{in-state:} \qquad F_0, F_1, F_2, F_3, F_4, F_5, F_6, F_7, F_8$$
$$\text{out-state:} \qquad F_0, F_1, F_4, F_3, F_2, F_7, F_8, F_5, F_6.$$

2. The *half-way wall bounceback*: Consider a channel with periodic boundary conditions in x-directions and walls at the lower and upper boundary.
   The first and last nodes in y-direction, i.e. with indices $j = 1$ and $j = N$,

are 'dry' (*dry* or *wall nodes*), i.e. outside the 'wet' domain which encompasses nodes with $j = 2$ to $j = N - 1$ (*wet* or *interior* or *fluid nodes*). The lower (upper) boundary is located half-way between the first and second (last but one and last) nodes in y-direction. Particle distributions are propagated between wet and dry nodes and vice versa. On the dry nodes no collision or forcing is performed and the distributions are all bounced back. Note that the width of the channel is one unit smaller than that with the complete bounceback scheme.

Plane Poiseuille flow: analytic solution. Only few analytical solutions of the Navier-Stokes equation are known. One is the plane Poiseuille flow in a channel of width $2L$ where the flow is steady ($\partial/\partial t = 0$), in x-direction $\boldsymbol{u} = (u, v) = (u(y), 0)$, with constant pressure (p and $\rho =$ constant) and without variations in x-direction ($\partial/\partial x = 0$). The flow is driven by a constant force $\boldsymbol{K} = K\boldsymbol{e}_x$. Accordingly the Navier-Stokes equation reduces to an ordinary differential equation for $u(y)$

$$\nu \frac{d^2 u}{dy^2} + K = 0$$

and the continuity equation $\dfrac{du(y)}{dx} = 0$ is satisfied. At the channel walls no-slip boundary conditions apply, i.e. $u(y) = 0$ at $y = -L$ and $y = L$. The analytical solution is a parabola

$$u(y) = \frac{K}{2\nu} \left(L^2 - y^2 \right). \tag{5.2.56}$$

Plane Poiseuille flow: numerical simulation. In the numerical simulation the velocity is initialized to zero and the mass density to a constant value of $\rho = 1$. In order to start the fluid flow a constant force has to be applied at every time step. Here the microscopic forcing method is applied.

First the simplest no-slip scheme, namely bounceback will be used. The result of an integration over a domain with 20 time 20 nodes is shown in Fig. 5.2.6 together with the analytical solution. The fluid is resting in the beginning and then is slowly accelarated. After $t = 1200$ time steps the mean x-momentum becomes steady (plot at lower right). The final velocity profile $u(y)$ (+ in plot at upper left) is compared with the analytical solution (Eq. 5.2.56) assuming a channel width $2L = YMAX - 1 = 19$ where $YMAX = 20$ is the number of nodes in y-direction. The numerical solution is zero on the boundary points (nodes $j = 1$ and $j = YMAX$ or $y = \pm 9.5$) and the profile looks similar to a parabola. The most striking difference compared to the analytical solution is the lower value of the maximum velocity. One can fit a parabola to the numerical data with equal values of the maximum velocity (plot at upper right). Deviations between this fit and the numerical values or most obvious near the boundaries. The plot at the lower left shows the numerical solution together with the analytical solution but the latter one now based on a smaller channel width $2L = YMAX - 2 = 18$. This solution compares very well with the analytical solution except for the two boundary nodes. This figure suggests the following interpretation. The boundary is located half-way between the first and second node respectively half-way between the second to last and the last node. Therefore the 'wet' channel has only a width of $2L = YMAX - 2 = 18$. The nodes at $j = 1$ and $j = YMAX$ are auxiliary nodes which take care of the correct boundary conditions at $j = 3/2$ and $j = YMAX - 1/2$. The velocity values on these auxiliary nodes should not be interpreted as flow velocities (these nodes are located 'on land' and could by called 'dry nodes').

Fig. 5.2.6. *No-slip boundary conditions and Poiseuille flow (see text for discussion).*

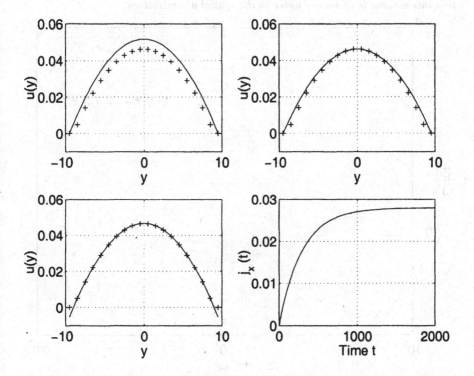

Numerical experiments with different channel widths $2L = 16, 32, 64, 128$ (different spatial resolutions of the parabola) have been performed while keeping constant the viscosity ν and the product of channel width and maximum speed (by varying the forcing). Thus the Reynolds number is kept constant. The Mach number varies but for $L \geq 16$ is small compared to 1. The error has been calculated by

$$\text{Error} := \frac{1}{N} \sqrt{\sum_{i=1}^{N} \left(\hat{u}_i^{(n)} - \hat{u}_i^{(a)}\right)^2}$$

where the summation is over $N = 14$ inner points and $\hat{u}_i^{(n)}$ and $\hat{u}_i^{(a)}$ are the normalized numerical and analytical solutions with maximum speed equal to one. Fig. (5.2.7) shows the error as a function of L. The slop of the curve is close to -2 which indicates that this scheme is of second order in the spatial discretization.

Fig. 5.2.7. *Error of the numerical compared to the analytical solution of the Poiseuille flow as a function of spatial resolution. The bounceback scheme is applied and the resulting data are interpreted such that the boundaries are located between the first and second respectively between the second to last and the last node (compare also Fig. 5.2.6C). The slop of the curve is close to -2 which indicates that this scheme is of second order in the spatial discretization.*

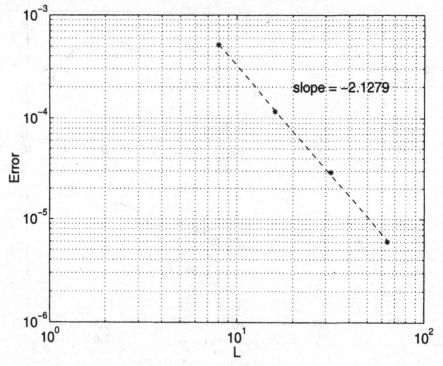

Slip boundary condition. To test slip boundary condition ($\dfrac{\partial u_t}{\partial x_n} = 0$ at $|y| \rightarrow \infty$, i.e. the normal derivative of the tangential velocity component u_t vanishes on the boundary) a shear layer with

$$u = \begin{cases} -U & \text{for } y < 0 \\ +U & \text{for } y > 0 \end{cases}, \quad v = 0$$

is initialized at $t = 0$. The analytical solution of the Navier-Stokes equation in $-\infty < y < \infty$ for this initial condition is known:

$$u(y,t) = U\,erf\left(\frac{y}{\sqrt{4\nu t}}\right), \quad v = 0$$

where $erf(arg)$ is the error function.

The slip conditions have been implemented by reflection of the distribution F_i on boundary nodes. The result of a numerical simulation ($U = 0.1$) in a channel of finite width ($2L = 18$) are shown together with the analytical solution in Fig. 5.2.8. The implementation of the simple slip scheme works very well. Small deviations between numerical and analytical solution are to be expected due to the finite width of the channel in the simulation.

Fig. 5.2.8. *Shear layer flow in a channel: test of slip boundary conditions (see text).*

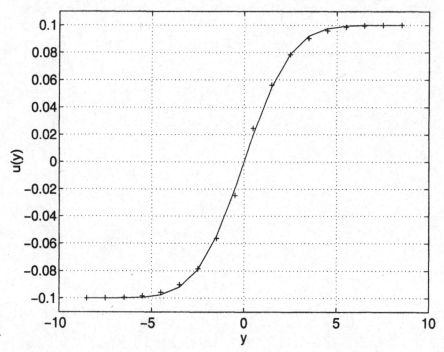

Further reading: boundary conditions for LBM
Skordos (1993), Ziegler (1993), Ginzbourg and Adler (1994), He and Zou

(1995), Noble et al. (1995a,b), Inamuro et al. (1995), Noble et al. (1996), Chen et al. (1996), Ginzbourg and d'Humiéres (1996), Maier et al. (1996), Filippova and Hänel (1997), Gallivan et al. (1997), Stockman et al. (1997), Kandhai et al. (1999).

5.3 Hydrodynamic lattice Boltzmann models in 3D

BGK-LBMs have been discussed so far only in two dimensions. The extension to three dimensions is straightforward. One just has to choose a 3D lattice and calculate appropriate equilibrium distributions. In this section we will calculate equilibrium distributions by the method of Koelman (1991) for 3D lattices with 19 and 15 velocities.

The derivation of the equilibrium distribution functions $F_i^{(eq)}$ (5.2.11) by minimizing the functional (5.2.18) is independent of the lattice velocities c_i and the dimension D. Only the constraints (5.2.6 to 5.2.8) for the velocity moments were used in Section (5.2). Thus one can use the form of the equilibrium distribution functions $F_i^{(eq)}$ (5.2.11) with the appropriate lattice velocities c_i and weights W_i.

5.3.1 3D-LBM with 19 velocities

For hydrodynamic simulation d'Humières et al. (1986) proposed a multi-speed lattice-gas cellular automata over a cubic lattice with 19 velocities (compare Section 3.3). This cubic lattice D3Q19 is defined by the following velocities

$$
\begin{aligned}
c_0 &= (0,0) \\
c_{1,2},\, c_{3,4},\, c_{5,6} &= (\pm 1, 0, 0),\, (0, \pm 1, 0)\,(0, 0, \pm 1) \\
c_{7,\ldots,10},\, c_{11,\ldots,14},\, c_{15,\ldots,18} &= (\pm 1, \pm 1, 0),\, (\pm 1, 0, \pm 1),\, (0, \pm 1, \pm 1).
\end{aligned}
$$

The calculation of the W_i for the 3D model proceeds in close analogy with the 2D case. There are three different speeds: 1 time speed 0 (rest particle), 6 times speed 1 (1-particles) and 12 times speed $\sqrt{2}$ ($\sqrt{2}$-particles). The even moments yield four independent equations for the calculation of W_0, W_1, W_2 and k_bT/m:

− 0. moment:

$$
\sum_i W_i = W_0 + 6W_1 + 12W_2 = \rho_0
$$

− 2. moment:

$$
\sum_i c_{\alpha x}^2 W_i = 2W_1 + 8W_2 = \rho_0 \frac{kT}{m}
$$

− 4. moment:

$$\sum_i c_{\alpha x}^4 W_i = 2W_1 + 8W_2 = 3\rho_0 \left(\frac{kT}{m}\right)^2$$

$$\sum_i c_{\alpha x}^2 c_{\alpha y}^2 W_i = 4W_2 = \rho_0 \left(\frac{kT}{m}\right)^2$$

The solution reads

$$W_0 = \frac{\rho_0}{3}$$

$$W_1 = \frac{\rho_0}{2} \left(\frac{kT}{m}\right)^2 = \frac{\rho_0}{18}$$

$$W_2 = \frac{\rho_0}{4} \left(\frac{kT}{m}\right)^2 = \frac{\rho_0}{36} \qquad (5.3.1)$$

$$\frac{kT}{m} = \frac{1}{3}.$$

The odd velocity moments over W_i vanish.

5.3.2 3D-LBM with 15 velocities and Koelman distribution

The lattice velocities of the D3Q15 lattice read

$$
\begin{aligned}
c_0 &= (0,0,0) & &\text{rest particle} \\
c_{1,2}, \, c_{3,4}, \, c_{5,6} &= (\pm 2,0,0), (0,\pm 2,0)\,(0,0,\pm 2) & &\text{2-particles} \\
c_{7,\ldots,14} &= (\pm 1,\pm 1,\pm 1) & &\sqrt{3}\text{-particles.}
\end{aligned}
$$

There are three different speeds: 1 time speed 0, 6 times speed 2 and 8 times speed $\sqrt{3}$. The even moments yield four independent equations for the calculation of W_0, W_2, W_3 and $k_b T/m$:

− 0. moment:

$$\sum_i W_i = W_0 + 6W_2 + 8W_3 = \rho_0$$

− 2. moment:

$$\sum_i c_{\alpha x}^2 W_i = 8W_2 + 8W_3 = \rho_0 \frac{kT}{m}$$

– 4. moment:

$$\sum_i c_{\alpha x}^4 W_i = 32W_2 + 8W_3 = 3\rho_0 \left(\frac{kT}{m}\right)^2$$

$$\sum_i c_{\alpha x}^2 c_{\alpha y}^2 W_i = 8W_3 = \rho_0 \left(\frac{kT}{m}\right)^2$$

The solution reads

$$W_0 = \frac{7}{18}\rho_0$$

$$W_2 = \frac{\rho_0}{16}\left(\frac{kT}{m}\right)^2 = \frac{\rho_0}{36}$$

$$W_3 = \frac{\rho_0}{8}\left(\frac{kT}{m}\right)^2 = \frac{\rho_0}{18}$$

$$\frac{kT}{m} = \frac{2}{3}.$$

The odd velocity moments over W_i vanish.

5.3.3 3D-LBM with 15 velocities proposed by Chen et al. (D3Q15)

The equilibrium distributions are not unique and other choices are possible. As an example we give the equilibrium distributions proposed by Chen et al. (1992):

$$F_0^{eq} = d^{(0)} + \delta^{(0)}v^2$$

$$F_i^{eq} = d^{(1)} + \beta^{(1)}c_i \cdot v + \gamma^{(1)}(c_i \cdot v)^2 + \delta^{(1)}v^2 \quad \alpha = 1, ..., 6$$

$$F_i^{eq} = d^{(2)} + \beta^{(2)}c_i \cdot v + \gamma^{(2)}(c_i \cdot v)^2 + \delta^{(2)}v^2 \quad \alpha = 7, ..., 14$$

where

$$d^{(0)} = d^{(1)} = \frac{\rho}{11}, \quad d^{(2)} = \frac{\rho}{22}$$

$$\alpha^{(1)} = \frac{\rho}{24}, \quad \alpha^{(2)} = \frac{\rho}{12}$$

$$\gamma^{(1)} = \frac{\rho}{32}, \quad \gamma^{(2)} = \frac{\rho}{16}$$

$$\delta^{(0)} = -\frac{7}{24}\rho, \quad \delta^{(1)} = -\frac{\rho}{48}, \quad \delta^{(2)} = -\frac{\rho}{24}.$$

These distributions will not be considered in the following.

5.4 Equilibrium distributions: the ansatz method

"The question that we are most often asked about cellular automata
is the following.
'I've been shown cellular automata that make surprisingly good models of, say, hydrodynamics, heat conduction, wave scattering, flow
through porous media, nucleation, dendritic growth, phase separation, etc. But I'm left with the impression that these are all *ad hoc*
models, arrived at by some sort of magic.'
'I'm a scientist, not a magician. Are there well-established *correspondence rules* that I can use to translate features of the system I want to
model into specifications for an adequate cellular-automaton model
of it?'
Physical modeling with cellular automata is a young discipline. Similar questions were certainly asked when differential equations were
new - and, despite three centuries of accumulated experience, modeling with differential equations still requires a bit of magic."
Toffoli und Margolus (1990)

*The problem addressed for CA by Toffoli and Margolus exists also for lattice
Boltzmann models. However, in what follows it will be shown that a LBM
for the Navier-Stokes equation can be developed by an almost straightforward method. In chapter 5.2 global equilibrium functions W_i for vanishing
velocity ($u = 0$) and constant density $\rho = \rho_0$ were determined in close analogy to the Maxwell distribution (more precise: the velocity moments of W_i
up to fourth order are equal to the corresponding velocity moments over
the Maxwell distribution; compare Eqs. 5.2.6 - 5.2.8). Subsequently the local
equilibrium distributions F_i have been derived using the maximum entropy
principle. These 'Koelman-distributions' lead to the Navier-Stokes equation
with isotropic advection, Galilean invariance and pressure which does not
explicitly depend on flow speed.*

*To be sure, these are not the only distributions over that lattice which in the
macroscopic limit yield the Navier-Stokes equation. This is not a problem
because the equilibrium distributions of our artificial microworld are not of
interest by themselves.*

*In the current section I will discuss an alternative approach to suitable equi-
librium distributions F_i whereby an ansatz for F_i will be used. After the
multiscale analysis the free parameters in the ansatz will be chosen such that
isotropy etc. are assured. This alternative approach will be of central importance for the development of lattice Boltzmann models for given differential
equations.*

To keep things simple the presentation will be restricted to a twodimensional model. The extension to 3D is straightforward. An LBM is defined by three ingredients:

1. A kinetic equation: here the meanwhile well-established BGK equation is chosen.

2. A lattice with sufficient symmetry: in order to compare results with those of the previous chapter the D2Q9 lattice is chosen; for a model based on D2Q7 see exercise (5.4.3).

3. Equilibrium distributions: see below.

Instead of using the maximum entropy principle to derive equilibrium distributions one may propose an ansatz with free parameters. How should the ansatz look like?

1. From lattice-gas cellular automata (FHP, FCHC, PI) and the Koelman model we know that terms quadratic in u and $c_i \cdot u$ in the equilibrium distributions yield the nonlinear advection term of the Navier-Stokes equation; no higher moments are required.

2. The free parameters of the ansatz should depend only on the mass density and the speeds $|c_i|$ but not on the directions of the c_i.

This suggests the following ansatz with ten free parameters:

$$
\begin{aligned}
F_i &= A_0 + D_0 u^2 & i &= 0 \\
F_i &= A_1 + B_1 c_i \cdot u + C_1 (c_i \cdot u)^2 + D_1 u^2 & i &= 1, 2, 3, 4 \quad (5.4.1) \\
F_i &= A_2 + B_2 c_i \cdot u + C_2 (c_i \cdot u)^2 + D_2 u^2 & i &= 5, 6, 7, 8.
\end{aligned}
$$

The definitions of mass and momentum densities give three scalar constraints

$$
\rho := \sum_i F_i = \underbrace{A_0 + 4(A_1 + A_2)}_{= \rho} + u^2 \underbrace{(D_0 + 4D_1 + 4D_2 + 2C_1 + 4C_2)}_{= 0} \quad (5.4.2)
$$

$$
j := \sum_i c_i F_i = \underbrace{(2B_1 + 4B_2)}_{= \rho} u. \quad (5.4.3)
$$

5.4.1 Multi-scale analysis

The multi-scale analysis proceeds as in Section 5.2.3 up to Eq. (5.2.31) inclusively. Not until the calculation of the momentum flux tensor $P_{\alpha\beta}^{(0)}$ the specific form of the equilibrium distribution has to be taken into account:

$$P^{(0)}_{\alpha\beta} := \sum_i c_{i\alpha} c_{i\beta} F^{(0)}_i$$

$$= \sum_{i=1}^{4} c_{i\alpha} c_{i\beta} \left[A_1 + B_1 c_i \cdot u + C_1 (c_i \cdot u)^2 + D_1 u^2 \right]$$

$$+ \sum_{i=5}^{8} c_{i\alpha} c_{i\beta} \left[A_2 + B_2 c_i \cdot u + C_2 (c_i \cdot u)^2 + D_2 u^2 \right]$$

$$= (2A_1 + 4A_2) \delta_{\alpha\beta} + \left(2C_1 u_\alpha^2 + 4C_2 u^2 \right) \delta_{\alpha\beta}$$

$$+ 8C_2 u_\alpha u_\beta (1 - \delta_{\alpha\beta}) + (2D_1 + 4D_2) u^2 \delta_{\alpha\beta}$$

with $u = (u, v) = (u_1, u_2)$. The goal is to transform the tensor

$$P^{(0)}_{\alpha\beta} = \begin{pmatrix} 2A_1 + 4A_2 + 2C_1 u^2 & 8C_2 uv \\ +(4C_2 + 2D_1 + 4D_2)u^2 & \\ & \\ 8C_2 uv & 2A_1 + 4A_2 + 2C_1 v^2 \\ & +(4C_2 + 2D_1 + 4D_2)u^2 \end{pmatrix}$$

into

$$\rho \begin{pmatrix} u^2 & uv \\ uv & v^2 \end{pmatrix} + p\, \delta_{\alpha\beta}.$$

The portion of the tensor which is independent of the flow velocity u yields the pressure p:

$$p = 2A_1 + 4A_2.$$

All other terms lead to the following constraints for the free parameters

$$4C_2 + 2D_1 + 4D_2 = 0 \quad (u^2 \text{ term in } P^{(0)}_{11} \text{ must vanish})$$

$$2C_1 = \rho \quad (u^2 \text{ term in } P^{(0)}_{11} \text{ must yield } \rho u^2)$$

$$8C_2 = \rho \quad (uv \text{ term in } P^{(0)}_{12} \text{ must yield } \rho uv)$$

and therefore

$$C_1 = \frac{\rho}{2} \quad \text{and} \quad C_2 = \frac{\rho}{8}.$$

The remaining eight unknowns (A_0, A_1, A_2, B_1, B_2, D_0, D_1, D_2) are constrained by only four linear equations

$$A_0 + 4(A_1 + A_2) = \rho$$
$$2B_1 + 4B_2 = \rho$$
$$D_0 + 4D_1 + 4D_2 = -2C_1 - 4C_2 = -\frac{3}{2}\rho$$
$$2D_1 + 4D_2 = -\frac{\rho}{2}.$$

Therefore in addition some arbitrary restrictions can be imposed. Here is just one possibility:

$$\frac{A_0}{A_1} = \frac{A_1}{A_2} = \frac{B_1}{B_2} = \frac{D_0}{D_1} =: r \tag{5.4.4}$$

where r is a free parameter. Thus the coefficients are functions of r:

$$A_0 = \frac{r^2}{(r+2)^2}\rho, \quad A_1 = \frac{r}{(r+2)^2}\rho, \quad A_2 = \frac{1}{(r+2)^2}\rho$$

$$B_1 = \frac{r}{4+2r}\rho, \quad B_2 = \frac{1}{4+2r}\rho$$

$$D_0 = -\frac{r}{2+r}\rho, \quad D_1 = -\frac{1}{2+r}\rho, \quad D_2 = -\frac{r-2}{16+8r}\rho$$

$$p = \frac{2}{r+2}\rho = c_s^2\rho \tag{5.4.5}$$

The speed of sound

$$c_s = \sqrt{\frac{dp}{d\rho}} = \sqrt{\frac{2}{r+2}}$$

is tunable by the parameter r (compare Fig. 5.4.1).

The additional restriction

$$\frac{D_1}{D_2} = r$$

leads to a quadratic equation for r with solutions $r_1 = 4$ and $r_2 = -2$. The solution r_2 yields negative equilibrium distribution even at small Mach numbers and an infinite pressure. The solution $r = 4$ (by the way: C_1/C_2 also equals 4) leads to

$$A_0 = \frac{4}{9}\rho, \quad A_1 = \frac{1}{9}\rho, \quad A_2 = \frac{1}{36}\rho, \quad B_1 = \frac{1}{3}\rho, \quad B_2 = \frac{1}{12}\rho,$$

$$D_0 = -\frac{2}{3}\rho, \quad D_1 = -\frac{1}{6}\rho, \quad D_2 = -\frac{1}{24}\rho, \quad p = \frac{1}{3}\rho = c_s^2\rho$$

with the speed of sound

$$c_s = \frac{1}{\sqrt{3}}.$$

The equilibrium distribution read

Fig. 5.4.1. *The speed of sound c_s as a function of the parameter r.*

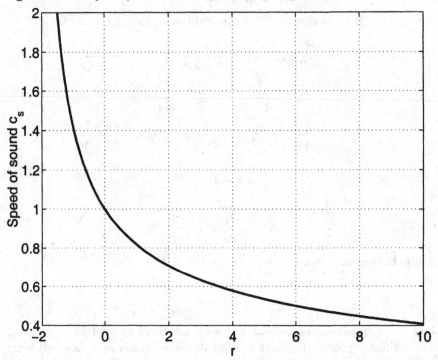

$$F_i = \frac{4}{9}\rho \left[1 - \frac{3}{2}u^2\right] \qquad\qquad i = 0$$

$$F_i = \frac{1}{9}\rho \left[1 + 3c_i \cdot u + \frac{9}{2}(c_i \cdot u)^2 - \frac{3}{2}u^2\right] \qquad i = 1, 2, 3, 4$$

$$F_i = \frac{1}{36}\rho \left[1 + 3c_i \cdot u + \frac{9}{2}(c_i \cdot u)^2 - \frac{3}{2}u^2\right] \qquad i = 5, 6, 7, 8.$$

The distribution functions are identical to those derived by the maximum entropy principle (compare Eq. 5.2.12 with $c = 1$)! These equilibrium distributions were also used by Martinez et al. (1994).

In summary, the equilibrium distribution have been derived in the current section by adjusting the free parameters of a plausible ansatz after the multi-scale analysis such that the desired macroscopic equations will result. For the model discussed above, there are not enough constraints to derive a unique solution for all free parameters. This freedom can be used, for example, to tune the speed of sound. The equilibrium distribution derived by the maximum entropy principle is a special case of the above calculated solutions.

The alternative approach to derive equilibrium distributions opens up the prospect to develop lattice Boltzmann models for other differential equations. From an ansatz for the equilibrium distributions and the conservation laws the desired properties of the macroscopic equations can be implemented into the microscopic model by appropriate choice of the free parameter after the multi-scale analysis. The ansatz method will be applied in the development of thermal LBM (Section 5.5) and LBM for the diffusion equation (Section 5.8).

5.4.2 Negative distribution functions at high speed of sound

Eq. (5.4.5) suggests that the speed of sound c_s might be tuned to arbitrary values by variing the parameter r. For small or even negative r, however, the distribution functions F_{si} become negative which leads to numerical instability very quickly. Let r_c be the critical r where one of the F_{si} first vanishes. It can be shown (see Exercise 5.4.2) that r_c depends on the velocity u as follows

$$r_c = \frac{2u^2}{1 - u^2} \qquad\qquad (5.4.6)$$

and therefore

$$c_s < c_s(r_c) = \sqrt{1 - u^2}. \qquad\qquad (5.4.7)$$

The Navier-Stokes equation has been derived in the small Mach number limit. Thus

$$M_a := \frac{|u|}{c_s} = \frac{|u|}{\sqrt{1 - u^2}} \qquad\qquad (5.4.8)$$

should remain small compared to 1. Consequently u^2 must remain small compared to 1 (singularity of M at $u^2 = 1$) and therefore $c_s < 1$ (compare Fig. 5.4.2).

Fig. 5.4.2. *The Mach number M_a and the speed of sound c_s as a function of the speed $|u|$.*

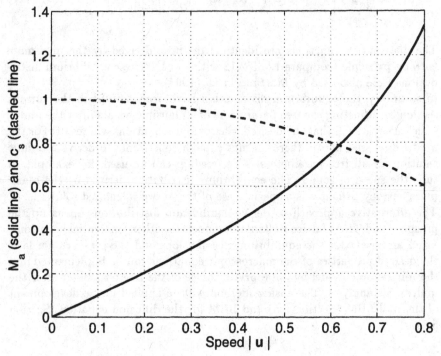

Exercise 5.4.1. (***)
Show that even when skipping the auxiliary constraints (5.4.4) while keeping $F_{si} > 0$ the speed of sound cannot exceed 1.

Exercise 5.4.2. (**)
Derive eq. (5.4.6).

Exercise 5.4.3. (***)
Calculate equilibrium distributions for an LBM for the Navier-Stokes equation in 2D over the D2Q7 lattice (FHP with rest particles). This model is called *pressure-corrected* LBM (PCLBM) because the pressure does not depend explicitly on the flow speed.

5.5 Hydrodynamic LBM with energy equation

Using the ansatz method described in Section 5.4 it is possible to develop hydrodynamic LBMs including an energy equation (so-called thermal models).

Alexander, Chen and Sterling (1993) proposed a thermal LBM over the D2Q13-FHP lattice (multi-speed FHP) with the following lattice velocities c_i (compare Fig. 3.3.8):

$$c_i = (0,0) \qquad\qquad\qquad i = 0$$

$$c_i = \left(\cos\frac{2\pi k}{6}, \sin\frac{2\pi k}{6}\right) \qquad i = 1,2,...,6; \quad k = i$$

$$c_i = 2\left(\cos\frac{2\pi k}{6}, \sin\frac{2\pi k}{6}\right) \qquad i = 7,8,...,12; \quad k = i-6$$

For the equilibrium distributions they made an ansatz including terms up to third order in the flow velocity u

$$F_i^{eq} = A_\sigma + B_\sigma c_i \cdot u + C_\sigma (c_i \cdot u)^2 + D_\sigma u^2 + E_\sigma (c_i \cdot u)^3 + G_\sigma (c_i \cdot u) u^2, \quad (5.5.1)$$

whereby the 14 free parameters A_σ to G_σ depend only on mass and internal energy density. The index σ refers to the square of the speed which is equal to zero, one, or two.
The BGK kinetic equation

$$F_i(x + c_i, t + 1) = (1 - \omega)F_i + \omega F_i^{eq} \qquad (5.5.2)$$

is applied.
Mass density (ρ), velocity (u) and internal energy (ε_I) are defined as follows

$$\rho = \sum_i F_i \qquad (5.5.3)$$

$$\rho u = \sum_i c_i F_i \qquad (5.5.4)$$

$$\rho \varepsilon_I = \sum_i \frac{(c_i - u)^2}{2} F_i. \qquad (5.5.5)$$

The desired form of the macroscopic equations

$$\frac{\partial \rho}{\partial t} + \frac{\partial}{\partial x_\alpha}(\rho u_\alpha) = 0 \qquad (5.5.6)$$

$$\rho\frac{\partial u_\alpha}{\partial t}+\rho u_\beta\frac{\partial u_\alpha}{\partial x_\beta}=-\frac{\partial p}{\partial x_\alpha}+\frac{\partial}{\partial x_\alpha}\left(\lambda\frac{\partial u_\gamma}{\partial x_\gamma}\right)+\frac{\partial}{\partial x_\beta}\left[\mu\left(\frac{\partial u_\beta}{\partial x_\alpha}+\frac{\partial u_\alpha}{\partial x_\beta}\right)\right]\quad(5.5.7)$$

$$\rho\frac{\partial\varepsilon_I}{\partial t}+\rho u_\beta\frac{\partial\varepsilon_I}{\partial x_\beta}=-p\frac{\partial u_\gamma}{\partial x_\gamma}+\frac{\partial}{\partial x_\beta}\left(\kappa\frac{\partial T}{\partial x_\beta}\right)+\mu\left(\frac{\partial u_\beta}{\partial x_\alpha}+\frac{\partial u_\alpha}{\partial x_\beta}\right)\frac{\partial u_\beta}{\partial x_\alpha}+\lambda\left(\frac{\partial u_\gamma}{\partial x_\gamma}\right)^2$$
$$(5.5.8)$$

is obtained by the following choice for the coefficients (see Appendix 6.4)

$$A_0=\rho\left(1-\frac{5}{2}\varepsilon_I+2\varepsilon_I^2\right),\quad A_1=\rho\frac{4}{9}\left(\varepsilon_I-\varepsilon_I^2\right),\quad A_2=\rho\frac{1}{36}\left(-\varepsilon_I+4\varepsilon_I^2\right)$$
$$(5.5.9)$$

$$B_1=\rho\frac{4}{9}\left(1-\varepsilon_I\right),\quad B_2=\rho\frac{1}{36}\left(-1+4\varepsilon_I\right)\qquad(5.5.10)$$

$$C_1=\rho\frac{4}{9}\left(2-3\varepsilon_I\right),\quad C_2=\rho\frac{1}{72}\left(-1+6\varepsilon_I\right)\qquad(5.5.11)$$

$$D_0=\rho\frac{1}{4}\left(-5+8\varepsilon_I\right),\quad D_1=\rho\frac{2}{9}\left(-1+\varepsilon_I\right),\quad D_2=\rho\frac{1}{72}\left(1-4\varepsilon_I\right)\quad(5.5.12)$$

$$E_1=-\rho\frac{4}{27},\quad E_2=\rho\frac{1}{108}\qquad(5.5.13)$$

$$G_1=G_2=0.\qquad(5.5.14)$$

Alexander et al. (1993) give no statement concerning the uniqueness of the solution. The shear viscosity μ and the thermal conductivity κ are given by

$$\mu=\rho\varepsilon_I\left(\frac{1}{\omega}-\frac{1}{2}\right)\qquad(5.5.15)$$

$$\kappa=2\rho\varepsilon_I\left(\frac{1}{\omega}-\frac{1}{2}\right)\qquad(5.5.16)$$

(the compressional viscosity λ vanishes). Thus the Prandtl number is

$$Pr:=\frac{\mu}{\kappa}=\frac{1}{2}.\qquad(5.5.17)$$

The pressure p reads $p=\rho\varepsilon_I$. The temperature T is proportional to the internal energy ε_I.

Exercise 5.5.1. (**)

Is it possible to define a thermal LBM over the D2Q9 lattice?

Exercise 5.5.2. (***)
Is it possible to choose the coefficients such that the Prandtl number is 1 or even becomes a tunable parameter?

Further reading:
Chen et al. (1995a,b) introduced an additional parameter in order to tune the Prandtl number. McNamara et al. (1995) discussed numerical instabilities of the models proposed by Alexander et al. (1993) and Chen et al. (1995a,b). Vahala et al. (1996) applied the model of Alexander et al. (1993) to free-decaying turbulence in 2D. McNamara et al. (1997) proposed a 3D thermal LBM with 27 velocities. They apply a Lax-Wendroff scheme in order to improve numerical stability. They performed simulations of Rayleigh-Bénard convection in 2D and compared the results with those of an explicit finite-difference (FD) solver. The computer times of LBM and FD are comparable whereas LBM requires significantly more memory and stability is significantly poorer. Thus they conclude that currently there is no potential advantage in using a thermal LBM over a conventional FD solver. Hu et al. (1997) consider the energy levels, ε_σ, of the three different speeds as free parameters and thereby are able to tune the ratio of specific heats, γ. Shan (1997) proposed a two-component LBGK in which temperature is advected as an passive scalar and simulated Rayleigh-Bénard convection. Pavlo et al. (1998a,b) investigate linear stability of thermal LBMs. The model of Alexander et al. (1993) has been revisited by Boghosian and Coveney (1998). They showed "that it is possible to achieve variable (albeit density-dependent) Prandtl number even within a single-relaxation-time lattice-BGK model". He et al. (1998), Vahala et al. (1998b).

5.6 Stability of lattice Boltzmann models

Lattice Boltzmann schemes do not have an H-theorem and therefore are subject to numerical instabilities (Sterling and Chen, 1996). Linear stability analysis has been performed for various LB models (D2Q7, D2Q9, D3Q15) and different background flows (homogeneous or shear flow) by Sterling and Chen (1996) and Worthing et al. (1997). The stability does not only depend on the background flow but also on the mass fraction parameters (α, β) of the equilibrium distributions and the grid size (stability decreases with increasing grid size).

5.6.1 Nonlinear stability analysis of uniform flows

The time evolution of BGK lattice Boltzmann models is described by the kinetic equation

$$F_m(x + c_m, t+1) = F_m(x,t) - \omega \left[F_m(x,t) - F_m^{(0)}(x,t) \right] \qquad (5.6.1)$$

Inspection of Eq. (5.6.1) shows that an initially uniform flow, in the sense

$$F_m(x, t_0) = F_m(t_0),$$

will remain uniform at all later times and thus

$$F_m(t+1) = F_m(t) - \omega \left[F_m(t) - F_m^{(0)}(t) \right].$$

Furthermore, mass and momentum density are conserved and retain their initial value

$$\rho(x,t) = \rho(t_0) = \rho_0, \quad j(x,t) = j(t_0) = j_0$$

and therefore

$$F_m(t+1) = F_m(t) - \omega \left[F_m(t) - F_m^{(0)}(\rho_0, j_0) \right].$$

Subtracting $F_m^{(0)}(\rho_0, j_0)$ on both sides leads to

$$\tilde{F}_m(t+1) = (1 - \omega)\tilde{F}_m(t)$$

with $\tilde{F}_m(t) = F_m(t) - F_m^{(0)}(\rho_0, j_0)$. The evolution is stable in the sense that the magnitude of $\tilde{F}_m$ does not increase with time ($|\tilde{F}_m(t+1)| \leq |\tilde{F}_m(t)|$) if

$$|1 - \omega| < 1 \quad \rightarrow \quad 0 < \omega < 2 \quad \text{or} \quad \frac{1}{\omega} = \tau > \frac{1}{2}$$

(compare Fig. 5.6.1).

Fig. 5.6.1. *Stability range of BGK models.*

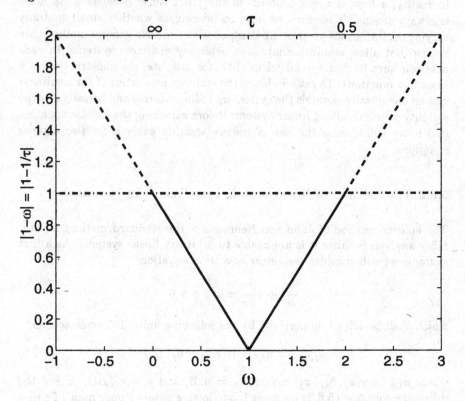

Exercise 5.6.1. (**)
Consider the kinetic equation

$$F_m(\pmb{x} + \pmb{c}_m, t + 1) = F_m(\pmb{x}, t) + \Omega_{ik}\left[F_k(\pmb{x}, t) - F_k^{(0)}(\pmb{x}, t)\right]$$

and show that for uniform flows the following equation holds

$$\tilde{\pmb{F}}(t + 1) = [\pmb{I} + \pmb{\Omega}]\tilde{\pmb{F}}(t)$$

where the vector $\tilde{\pmb{F}}(t)$ has components $\tilde{F}_m(t) = F_m(t) - F_m^{(0)}(\rho_0, \pmb{j}_0)$.

In reality, a flow is never uniform in the strict sense because some noise is always around. Therefore we have to investigate whether small spatially varying perturbations grow or are damped with time. No general methods are known that allow stability analysis of arbitrary nonlinear systems. In each case you have to find special tricks (like, for instance, the construction of a Liapunov function). To get an idea of the stability properties of the nonlinear system one usually expands the system up to linear terms and investigates the stability of the resulting linear system. Before attacking the kinetic equation (5.6.1) we will discuss the *von Neumann stability analysis* for two simpler equations.

5.6.2 The method of linear stability analysis (von Neumann)

The Fourier method of John von Neumann is the standard method for stability analysis because it is applicable to arbitrary linear systems. As a first example we will consider the linear advection equation

$$\frac{\partial u}{\partial t} + c\frac{\partial u}{\partial x} = 0, \quad c > 0$$

which shall be solved numerically by the following finite difference scheme

$$u_{j,n+1} = u_{j,n} - \mu\left(u_{j,n} - u_{j-1,n}\right) \tag{5.6.2}$$

where $u_{j,n} = u(x_j, t_n)$, $x_j = j\Delta x$, $t_n = n\Delta t$, and $\mu = c \cdot \Delta t/\Delta x$. For the difference equation (5.6.2) we make the following ansatz ('poor man's Fourier transform')

$$u_{j,n} = U_n e^{ikx} = U_n e^{ikj\Delta x} \tag{5.6.3}$$

(the actual solution is the real part) where k is the wave number of the spatial perturbation. Inserting of (5.6.3) into (5.6.2) leads to

$$U_{n+1} = (1 - \mu)U_n + \mu U_n e^{-ik\Delta x} = \lambda U_n$$

with

$$\lambda := 1 - \mu + \mu e^{-ik\Delta x}.$$

It follows that
$$|U_{n+1}| = |\lambda| \cdot |U_n| = |\lambda|^n \cdot |U_1|.$$
Stability in the sense that $|U_{n+1}|$ is bounded (i.e. $\lim_{n\to\infty} |U_{n+1}| < B < \infty$) yields the constraint
$$|\lambda| \leq 1.$$
Here we have
$$|\lambda|^2 = 1 - 2\mu(1 - \mu)(1 - \cos k\Delta x)$$
which is ≤ 1 when
$$0 \leq \mu = c\frac{\Delta t}{\Delta x} \leq 1$$
or
$$\frac{\Delta x}{\Delta t} \geq c \tag{5.6.4}$$
or
$$\boxed{0 \leq \Delta t \leq \frac{\Delta x}{c}.} \tag{5.6.5}$$

This is the stability condition we were looking for. According to Eq. (5.6.4) the 'grid speed' $c_{grid} := \Delta x/\Delta t$ has to be as fast or faster than the advection speed c (this holds in very similar form also for several spatial dimensions, for other explicit numerical schemes and other hyperbolic equations whereby the advection speed is eventually replaced by the speed of the fastes wave of the system).

As a second example we consider the diffusion equation
$$\frac{\partial T}{\partial t} = \kappa \frac{\partial^2 T}{\partial x^2}.$$
A finite difference approximation reads
$$T_{j,n+1} = T_{j,n} + \kappa \frac{\Delta t}{(\Delta x)^2} \left(T_{j+1,n} - 2T_{j,n} + T_{j-1,n} \right).$$
Inserting the ansatz
$$T_{j,n} = A_n e^{ikj\Delta x}$$
leads to
$$A_{n+1} = A_n + \mu' A_n \left(e^{ik\Delta x} - 2 + e^{-ik\Delta x} \right) = \lambda' A_n$$
with
$$\mu' = \kappa \frac{\Delta t}{(\Delta x)^2}$$
and
$$\lambda' = 1 - 2\mu'(1 - \cos k\Delta x)$$
The condition $|\lambda'| \leq 1$ yields the stability condition
$$0 \leq \mu' = \kappa \frac{\Delta t}{(\Delta x)^2} \leq \frac{1}{2}$$
or
$$\boxed{0 \leq \Delta t \leq \frac{(\Delta x)^2}{2\kappa}.} \tag{5.6.6}$$

5.6.3 Linear stability analysis of BGK lattice Boltzmann models

The following discussion is based on Worthing et al. (1997). The kinetic equation including external body forces $K_m(\boldsymbol{x}, t)$ reads

$$F_m(\boldsymbol{x} + \boldsymbol{c}_m, t+1) = F_m(\boldsymbol{x}, t) - \omega \left[F_m(\boldsymbol{x}, t) - F_m^{(0)}(\boldsymbol{x}, t) \right] + K_m(\boldsymbol{x}, t) \quad (5.6.7)$$

where F_m are the distribution functions and $F_m^{(0)}$ are the equilibrium distribution functions. Expansion about time-independent but otherwise arbitrary distribution functions $F_m^{(bf)}$ ('background flow') leads to

$$
\begin{aligned}
F_m(\boldsymbol{x}, t) &= F_m^{(bf)}(\boldsymbol{x}) + f_m(\boldsymbol{x}, t) \\
F_m(\boldsymbol{x} + \boldsymbol{c}_m, t+1) &= F_m^{(bf)}(\boldsymbol{x} + \boldsymbol{c}_m) + f_m(\boldsymbol{x} + \boldsymbol{c}_m, t+1) \\
F_m^{(0)}(\boldsymbol{x}, t) &= F_m^{(0)}(\{F_s(\boldsymbol{x}, t)\}) \\
&\approx F_m^{(0)}\left(\{F_s^{(bf)}(\boldsymbol{x})\}\right) + \sum_n \left(\frac{\partial F_m^{(0)}}{\partial F_n}\right)_{\{F_s^{(bf)}(\boldsymbol{x})\}} f_n(\boldsymbol{x}, t) \\
K_m(\boldsymbol{x}, t) &= K_m(\{F_s(\boldsymbol{x}, t)\}) \\
&\approx K_m\left(\{F_s^{(bf)}(\boldsymbol{x})\}\right) + \sum_n \left(\frac{\partial K_m}{\partial F_n}\right)_{\{F_s^{(bf)}(\boldsymbol{x})\}} f_n(\boldsymbol{x}, t)
\end{aligned}
$$

Inserting the expansions into the kinetic equation (5.6.7) results in

$$f_m(\boldsymbol{x} + \boldsymbol{c}_m, t+1) = G_m(\boldsymbol{x}) + (1 - \omega) f_m(\boldsymbol{x}, t) + \sum_n J_{mn}(\boldsymbol{x}) f_n(\boldsymbol{x}, t)$$

where

$$
\begin{aligned}
G_m(\boldsymbol{x}) &= F_m^{(bf)}(\boldsymbol{x}) - F_m^{(bf)}(\boldsymbol{x} + \boldsymbol{c}_m) - \omega \left[F_m^{(bf)}(\boldsymbol{x}) - F_m^{(0)}\left(\{F_s^{(bf)}(\boldsymbol{x})\}\right) \right] \\
&\quad + K_m\left(F_m^{(bf)}(\boldsymbol{x})\right)
\end{aligned}
$$

and

$$J_{mn} = \left[\omega \frac{\partial F_m^{(0)}}{\partial F_n} + \frac{\partial K_m}{\partial F_n} \right]_{\{F_s^{(bf)}(\boldsymbol{x})\}} . \quad (5.6.8)$$

Here we are interested only in instabilities with exponential growth. Therefore we will neglect the time-independent term $G_m(\boldsymbol{x})$ which can lead to linear growth at most. The rectangular domain of length L and width W comprises a lattice with nodes at $(x_r = 0, 1, 2, ..., L; y_s = 0, 1, 2, ..., W)$. We will now Fourier transform the linear equation

$$f_m(\boldsymbol{x} + \boldsymbol{c}_m, t + 1) = (1 - \omega)f_m(\boldsymbol{x}, t) + \sum_n J_{mn}(\boldsymbol{x})f_n(\boldsymbol{x}, t). \qquad (5.6.9)$$

The functions $f_m^{(k,l)}(t)$ and $f_m(x, y, t)$ are connected by the Fourier transform

$$f_m(x, y, t) = \sum_{k,l} f_m^{(k,l)}(t)e^{ikx2\pi/L}e^{ily2\pi/W}. \qquad (5.6.10)$$

Consequently one obtains

$$f_m(x + c_{x,m}, y + c_{y,m}, t + 1)$$
$$= \sum_{k,l} f_m^{(k,l)}(t + 1)e^{ik(x+c_{x,m})2\pi/L}e^{il(y+c_{y,m})2\pi/W}$$
$$= \sum_{k,l} f_m^{(k,l)}(t + 1)e^{ikx2\pi/L}e^{ily2\pi/W}e^{ikc_{x,m}2\pi/L}e^{ilc_{y,m}2\pi/W}$$

The analysis is further simplified by the assumption that the background flow (a shear layer or jet stream, for example) as well as the external forces depend only on y:

$$J_{mn}(y) = \sum_q J_{mn}^{(q)}e^{iqy2\pi/W}.$$

Inserting of the Fourier transforms into Eq. (5.6.9) one obtains

$$e^{ikc_{x,m}2\pi/L}e^{ilc_{y,m}2\pi/W}f_m^{(k,l)}(t+1) = (1-\omega)f_m^{(k,l)}(t) + \sum_{n,q} J_{mn}^{(q)}f_n^{(k,l-q)}(t)$$

Since the x modes remain uncoupled, they are considered independently via

$$e^{ikc_{x,m}2\pi/L}e^{ilc_{y,m}2\pi/W}f_m^{(l)}(t+1) = (1-\omega)f_m^{(l)}(t) + \sum_{n,q} J_{mn}^{(q)}f_n^{(l-q)}(t) \quad (5.6.11)$$

It can be shown (for some fine points see Worthing et al., 1997) that Eq. (5.6.11) can be written as a matrix iteration

$$\boldsymbol{f}_{t+1} = \mathcal{A}\boldsymbol{f}_t \qquad (5.6.12)$$

where the vector $\boldsymbol{f}$ has components $f_m^{(l)}$. If the spectral radius (= magnitude of the largest eigenvalue) of $\mathcal{A}$, $\rho(\mathcal{A})$, is larger than unity, then the system is said to be linear unstable (this corresponds to the case $|\lambda| > 1$ in the two examples with one component each discussed in Subsection 5.6.2). In case of uniform background flow and no external forcing considered by Sterling and Chen (1996) the matrix $\mathcal{A}$ reads

$$\boxed{\mathcal{A} = \boldsymbol{D}\left[(1-\omega)\boldsymbol{I} + \boldsymbol{J}\right]} \qquad (5.6.13)$$

where $\boldsymbol{D}$ is a diagonal matrix with components

$$D_{mn} = e^{-ikc_{x,m}2\pi/L} e^{-ilc_{y,m}2\pi/W} \delta_{mn},\qquad(5.6.14)$$

I is the identity matrix (diagonal matrix with value 1 of all diagonal components) and the components of J are given by

$$J_{mn} = \omega\frac{\partial F_m^{(0)}}{\partial F_n}.\qquad(5.6.15)$$

The matrix $\mathcal{A}$ (and therefore also its spectral radius $\rho(\mathcal{A})$) depends on the relative wave numbers ($\theta_x = 2\pi k/L$, $\theta_y = 2\pi l/W$), the uniform inital velocity (U), the collision parameter (ω) and the rest mass parameters (α, β; see below). Instability occurs if the maximal spectral radius (for given values of U, ω, α and β) becomes larger than 1 for any wave number:

$$\max_{\theta_x,\theta_y} \rho\left(\mathcal{A}\left[\theta_x,\theta_y,U,\omega,\alpha,\beta\right]\right) > 1.\qquad(5.6.16)$$

The components of J may be calculated as follows. The equilibrium distributions for the models D2Q7, D2Q9 and D3Q15 are all of the form

$$F_m^{(0)} = \rho\left[A_m + B_m c_m \cdot u + C_m (c_m \cdot u)^2 + D_m u^2\right]$$

with constants $A_m, ..., D_m$. The derivatives of mass and momentum density with respect to F_n are easy to calculate:

$$\rho = \sum_n F_n, \quad\rightarrow\quad \frac{\partial\rho}{\partial F_m} = 1,$$

$$j = \rho u = \sum_n F_n c_n, \quad\rightarrow\quad \frac{\partial j}{\partial F_n} = c_n$$

and consequently

$$\rho u^2 = \frac{j^2}{\rho}, \quad\rightarrow\quad \frac{\partial(\rho u^2)}{\partial F_n} = \frac{2j\dfrac{\partial j}{\partial F_n}\rho - \dfrac{\partial\rho}{\partial F_n}j^2}{\rho^2} = 2c_n \cdot u - u^2.$$

Finally we obtain

$$\frac{\partial F_m^{(0)}}{\partial F_n} = A_m + B_m c_m \cdot c_n + C_m\left[2(c_m \cdot c_n)(c_m \cdot u) - (c_m \cdot u)^2\right]$$
$$+ D_m\left[2c_n \cdot u - u^2\right].$$

The coefficients for the various models are listed below. The coefficients A_m include free parameters α and β (denoted as *rest mass parameters*) which can be tuned in order to improve stability.

FHP (D2Q7).

$$A_m = \alpha, \qquad B_m = 0, \quad C_m = 0, \quad D_m = -1; \quad m = 0$$
$$A_m = \frac{1-\alpha}{6}, \quad B_m = \tfrac{1}{3}, \quad C_m = \tfrac{2}{3}, \quad D_m = -\tfrac{1}{6}; \quad m = 1, ..., 6.$$

D2Q9.

$$A_m = \alpha, \qquad\qquad B_m = 0, \quad C_m = 0, \quad D_m = -\tfrac{2}{3}; \quad m = 0$$
$$A_m = \beta, \qquad\qquad B_m = \tfrac{1}{3}, \quad C_m = \tfrac{1}{2}, \quad D_m = -\tfrac{1}{6}; \quad m = 1, ..., 4$$
$$A_m = \frac{1-\alpha-4\beta}{4}, \quad B_m = \tfrac{1}{12}, \quad C_m = \tfrac{1}{8}, \quad D_m = -\tfrac{1}{24}; \quad m = 5, ..., 8$$

D3Q15.

$$m = 0:$$
$$A_m = \alpha, \qquad\qquad B_m = 0, \qquad C_m = 0, \qquad D_m = -\tfrac{1}{2}.$$
$$m = 1, ..., 6:$$
$$A_m = \beta, \qquad\qquad B_m = \tfrac{1}{3}, \qquad C_m = \tfrac{1}{2}, \qquad D_m = -\tfrac{1}{6}.$$
$$m = 7, ..., 14:$$
$$A_m = \frac{1-6\beta-\alpha}{8}, \qquad B_m = \tfrac{1}{24}, \qquad C_m = \tfrac{1}{16}, \qquad D_m = -\tfrac{1}{48}.$$

The eigenvalues of $\mathcal{A}$ have been calculated with a standard routine form MATLAB. The maximum eigenvalues for the D2Q7 (FHP) model with 1.) $(u, v) = (0.2, 0)$, $\alpha = 0.2$, $\theta_y = 0$ (solid line) and 2.) $(u, v) = (0.23, 0)$, $\alpha = 0.3$, $\theta_y = 0$ (broken line) are shown in Fig. 5.6.2 as a function of the relative wave number θ_x (the figure is identical to Figure 1 in Sterling and Chen (1996); this problem served as a test of the MATLAB script). For certain wave numbers the maximum eigenvalues become larger than unity and therefore the model is linear unstable for this choice of parameters. Detailed results on stability boundaries depending on the model parameters can be found in Sterling and Chen (1996) and Worthing et al. (1997).

5.6.4 Summary

The main results of Sterling and Chen (1996) and Worthing et al. (1997) can be summarized as follows:

- For the D2Q7 (FHP) model and homogeneous flow the wavenumber vector k of the most unstable mode is parallel to the velocity u.

Fig. 5.6.2. *The maximum eigenvalues for the D2Q7 (FHP) model with 1.) $(u, v) = (0.2, 0)$, $\alpha = 0.2$, $\theta_y = 0$ (solid line) and 2.) $(u, v) = (0.23, 0)$, $\alpha = 0.3$, $\theta_y = 0$ (broken line) are shown as a function of the relative wave number θ_x (compare Fig. 1 in Sterling and Chen, 1996).*

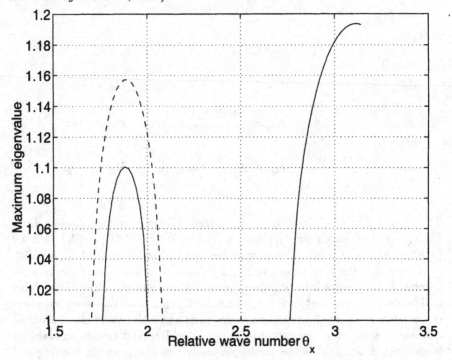

- For the D2Q9 model and homogeneous flow the wavenumber $|\boldsymbol{k}|$ of the most unstable mode is equal to about 2.3 ($\boldsymbol{k}$ is not necessary parallel to $\boldsymbol{u}$). An explanation for the 'magic' value 2.3 is not known.

- The stability domain as a function of the rest mass parameters α and β of the D2Q9 model shrinks with increasing lattice size. The 'canonical' values $\alpha = 4/9$, $\beta = 1/9$, which have been derived from maximum entropy principle (compare Section 5.2), lie inside one of the stability islands (compare Fig. 5.6.3).

- The D2Q7 model is less stable than the D2Q9 model in the sense that instability occurs already at smaller flow velocities (compare Fig. 2 in Sterling and Chen, 1996).

- The D2Q9 lattice is a projection of the D3Q15 lattice. Therefore it may be come at no big surprise that their linear stability properties for homogeneous background flows show some similarities. The wavenumber $|\boldsymbol{k}|$ of the most unstable mode is again equal to about 2.3 (Sterling and Chen, 1996).

- The stability domain shrinks further when the background flow includes shear. Please note that the usual linear stability analysis assumes a time-independent background flow whereas free shear layers decay by momentum diffusion. Therefore predictions from linear stability analysis are less reliable at high viscosities where the shear flow may decay before an instability has enough time to develop. Thus the unstable region is smaller than predicted (see, for example, Fig. 9 in Worthing et al., 1997). However, the linear stability analysis works fine in the low viscosity region (high Reynolds numbers) which is of most interest.

To the best of our knowledge linear stability analysis for D3Q19 and for thermal LB models is not available yet.

Exercise 5.6.2. (***)
Propose equilibrium distributions with rest mass parameters α and β for the D3Q19 model, investigate linear stability and compare the results with the linear stability properties of D3Q15.

Fig. 5.6.3. *Contour plot of the spectral radius as a function of the rest mass parameters α and β for the D2Q9 model. On a lattice with 64 times 64 grid points the stability region shrinks to small islands (domains inside the contour lines). The 'canonical' values $\alpha = 4/9$, $\beta = 1/9$, which have been derived from maximum entropy principle, lie inside one of the stability islands.*

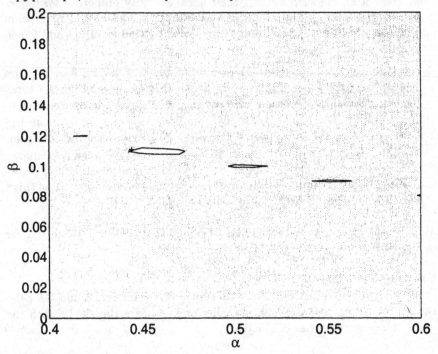

5.7 Simulating ocean circulation with LBM

Additional forces - wind stress, Coriolis force, and frictional forces - were implemented into a LB model with 9 lattice velocities in two dimensions (D2Q9). With this extended model it is possible to simulate the wind-driven circulation of a barotropic ocean. Model results are compared with analytical solutions of the linearized problem by Munk and with a finite difference model of the full nonlinear problem. The implementation of various boundary conditions and of body forces is discussed in some detail.

5.7.1 Introduction

Despite the rapid development of computer technology during the last decades, simulation of global ocean circulation is still limited by computer resources because of several small scale processes which are relevant for large scale features. Increasing computer power will not solve these problems during the next few years. More promising is the development of new methods like 'active nesting' (Spall and Holland, 1991, Fox and Maskell, 1995 and 1996) to deal with small scale dynamics in selected areas or the application of other numerical methods like finite elements or LB models.

Because of the strict locality of the 'collisions' in LB models they are especially well suited for massive parallel computers. The code is extremely simple compared to typical ocean circulation models like MOM (Pacanowski et al., 1991). No elliptic equation has to be solved. Here the simplest case will be considered, namely the wind-driven circulation of a barotropic ocean in a rectangular domain. The main goal of this work is to test whether the LB model yields results which are in quantitative agreement with the analytical solution of the linearized problem and with the numerical (finite differences, finite volumes)) solution of the nonlinear problem.

5.7.2 The model of Munk (1950)

One of the grand challenges of physical oceanography in the first half of the 20th century was the explanation of the western boundary currents like the Gulf Stream, the Agulhas or the Kuroshio. Ekman (1905 and 1923), Sverdrup (1947) and others had made important contributions to the theory of wind-driven circulation but could not explain the intensification of the flow near the western boundaries of oceanic basins. In the late 40ties in the time span of only two years different approaches for the basin-wide circulation were proposed by Stommel (1948) and Munk (1950) .

With the North Atlantic in mind Munk defined the following problem. Consider a rectangular ($L \times H$) flat-bottom barotropic (vertically integrated) ocean which is driven by wind stress of the form (locally cartesian coordinates x, y where x is eastward and y northward)

$$T_x = -T_0 \cos\left(\frac{\pi}{H}y\right) \quad \text{and} \quad T_y = 0, \tag{5.7.1}$$

corresponding to westerly wind in mid latitudes and easterly wind in low latitudes. The Navier-Stokes equation contains only molecular diffusion as dissipative process. For large scale oceanic circulation molecular diffusion does not play a role. To get rid of the vorticity imparted by the wind Munk replaced the molecular viscosity coefficient, ν, by the so-called *eddy viscosity coefficient*, A, which is several orders of magnitude larger than ν. The *Laplacian friction*, $A\nabla^2 u$, can be interpreted as a simple parameterization of subscale processes. The equation of motion thus reads

$$\frac{\partial u}{\partial t} + (u\nabla)\,u + f\hat{u} + \frac{1}{\rho}\nabla p - A\nabla^2 u - T = 0 \tag{5.7.2}$$

with $\hat{u} = (-v, u)$. The Coriolis parameter f is approximated by $f \simeq f_0 + \beta y$ (β-plane), where

$$f_0 = 2\Omega \sin\varphi_0 \quad \text{and} \quad \beta = \left(\frac{\partial f}{\partial y}\right)_{\varphi_0} = \frac{2\Omega \cos\varphi_0}{R}. \tag{5.7.3}$$

$R = 6371$ km is the mean radius of the Earth, $\Omega = 7.29\cdot 10^{-5}$ s^{-1} the angular velocity of the Earth and φ_0 the reference latitude. $A \approx 10^4$ m^2 s^{-1} is the horizontal eddy viscosity coefficient. This value corresponds to a typical value of $200 - 250$ km for the widths of the western boundary currents (see Munk, 1950). The velocity u of an incompressible ($\nabla \cdot u = 0$) two-dimensional flow can be calculated from a *stream function* $\psi(x, y)$

$$u = -\frac{\partial \psi}{\partial y} \quad \text{and} \quad v = \frac{\partial \psi}{\partial x}. \tag{5.7.4}$$

Taking the curl of Eq. (5.7.2) to eliminate the pressure gradient one obtains the *vorticity equation*

$$\frac{\partial}{\partial t}\nabla^2\psi + J(\psi, \nabla^2\psi) + \beta\frac{\partial \psi}{\partial x} - A\nabla^4\psi + \left(\frac{\partial T_y}{\partial x} - \frac{\partial T_x}{\partial y}\right) = 0 \tag{5.7.5}$$

where

$$J(a, b) = \frac{\partial a}{\partial x}\frac{\partial b}{\partial y} - \frac{\partial a}{\partial y}\frac{\partial b}{\partial x}$$

is the *Jacobi operator*.

The analytical solution of the linear Munk problem. For the stationary and linear case the vorticity equation (5.7.5) simplifies:

$$\nabla^4\psi - \frac{\beta}{A}\frac{\partial\psi}{\partial x} = -\frac{T_0}{A}\frac{\pi}{H}\sin\left(\frac{\pi}{H}y\right) \qquad (5.7.6)$$

The characteristic length (*Munk scale*)

$$W_M = \left(\frac{A}{\beta}\right)^{1/3} \qquad (5.7.7)$$

gives an estimate of the width of the western boundary current (see below). In order to resolve this current in the numerical simulation the grid spacing, $c_i\Delta t$, has to be smaller than W_M (= 80 km for $\beta = 2\cdot10^{-11}$ m^{-1} s^{-1} and $A = 10^4$ m^2 s^{-1}).

Ideally one would like to derive an analytical solution with noslip conditions ($u = 0$) on all boundaries (an ocean basin bounded by continents on all sides). In terms of the streamfunction ψ this would mean that $\psi = 0$ on all boundaries, $\partial\psi/\partial x = 0$ on the west and east boundary, and $\partial\psi/\partial y = 0$ on the south and north boundary. These conditions cannot easily be fulfilled exactly (consider the Fourier expansion of the meriodinal variation of the zonal wind stress: for each term Y_n of the series a solution X_n of the differential equation has to be constructed and the sum of all these terms has to fulfill the boundary conditions). Thus we will somewhat relax the constrains and allow slip conditions ($\psi = 0$, $\frac{\partial^2\psi}{\partial y^2}$) at the southern and northern boundaries (these boundary conditions are appropriate for an ocean gyre bounded on the west and east by continents, and on the south and north by other gyres circulating in the opposite direction).

The exact solution of Eq. (5.7.6) reads

$$\psi_{M,e} = -\frac{T_0\,H^3}{A\,\pi^3}\sin\left(\frac{\pi}{H}y\right)$$

$$(5.7.8)$$

$$\cdot\left\{1 + e^{\alpha_r kx}\left[p_1\cos\left(\alpha_i kx\right) - p_2\sin\left(\alpha_i kx\right)\right] + p_3 e^{\alpha_3 kx} + p_4 e^{\alpha_4 kx}\right\}$$

where $\alpha_{1,2} = \alpha_r \pm i\alpha_i$, α_3, and α_4 are roots of the characteristic equation

$$\left(\alpha^2 - \gamma^2\right)^2 = \alpha \qquad (5.7.9)$$

and the coefficients p_1 to p_4 are derived from the boundary conditions. The α_n and p_n have to be calculated numerically (Exercise 5.7.3).

Exercise 5.7.1. (**)
Derive the exact solution (Eq. 5.7.8) and the characteristic Eq. (5.7.9) of the linear Munk model (hint: use a separation ansatz).

Exercise 5.7.2. (*)
For small γ find approximate solutions of the characterstic Eq. (5.7.9).

Exercise 5.7.3. (**)
Calculate the α_m and p_m $(m = 1, ..., 4)$ for $A = 10^3$ m^2 s^{-1}, mean latitude $\phi_0 = 30°$, $L = H = 2 \cdot 10^6$ m.

5.7.3 The lattice Boltzmann model

We apply the lattice Boltzmann model introduced in Section 5.2, i.e. the BGK kinetic equation (Eq. 5.2.9) over the D2Q9 lattice with the equilibrium distributions given in Eq. (5.2.12). In the simulations of the linear Munk problem the nonlinear terms of the distribution functions have been dropped:

$$F_i = \frac{4}{9}\rho \qquad\qquad\qquad i = 0$$

$$F_i = \frac{1}{9}\rho \left[1 + 3\frac{c_i \cdot u}{c^2}\right] \qquad i = 1, 2, 3, 4$$

$$F_i = \frac{1}{36}\rho \left[1 + 3\frac{c_i \cdot u}{c^2}\right] \qquad i = 5, 6, 7, 8.$$

Here we will discuss some details of the coding of propagation, boundary conditions and forcing.

The grid consists of XMAX × YMAX sites where (XMAX-2) times (YMAX-2) points are 'wet' and the other sites are 'dry', i.e. they are located outside the domain. The boundary between land and ocean lies halfway between the dry and the neighboring wet site (Fig. 5.7.1).

Fig. 5.7.1. *Ocean model: lattice, wet (inside the dashed box) and dry (outside the dashed box) sites, slip or no-slip boundary conditions.*

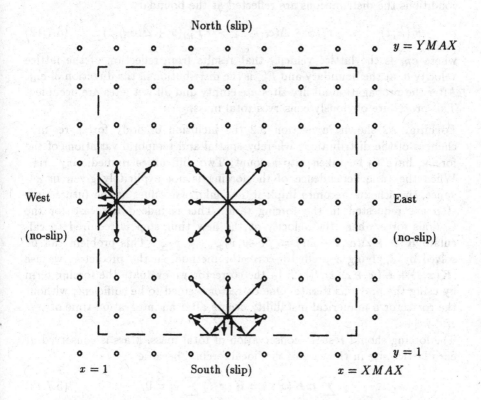

Propagation and boundary conditions. In the propagation step all distributions $F_i(x,t)$ (except the 'rest' distribution $F_0(x,t)$) of wet points proceed along the corresponding link (direction c_i) to the neighbor site and are stored in a second array, say F_i':

$$F_i(x,t) \quad \Rightarrow \quad F_i'(x + \Delta t c_i, t). \tag{5.7.10}$$

Near the boundary on wet sites not all F_i' are occupied because no distributions are propagated from dry sites. On the other hand some distributions from sites near the boundary make it to dry sites. The latter distributions propagate back to wet sites according to the local boundary conditions, i.e. in case of no-slip boundary conditions the F_i' bounce back:

$$F_i(x,t) \quad \Rightarrow \quad F_i'(x + \Delta t c_i, t) \quad \Rightarrow \quad F_{Ii}'(x,t) \tag{5.7.11}$$

where F_{Ii}' is the distribution in the direction of $-c_i$; in case of slip boundary conditions the distributions are reflected at the boundary:

$$F_i(x,t) \quad \Rightarrow \quad F_i'(x + \Delta t c_i, t) \quad \Rightarrow \quad F_{Ri}'(x + \Delta t c_{Ri}, t) \tag{5.7.12}$$

where c_{Ri} is the lattice velocity that results from reflection of the lattice velocity c_i at the boundary and F_{Ri}' is the distribution in the direction of c_{Ri}. After the propagation all dry sites are empty and all wet sites are occupied. This procedure obviously conserves total mass.

Forcing. As shown in Section 5.2 the inclusion of body forces requires changes of the distributions whereby spatial and temporal variations of the forcing have to be taken into account. Two difficulties immediately arise. When the time dependence of the forcing is not explicitly given in advance, the scheme becomes implicit, i.e. unknown values at the future time step are requested in the forcing term. This is indeed the case for the Coriolis force where the velocity at the new time step is required to calculate $K(x + \Delta t c_i, t + \Delta t) = f(-u, v)_{x + \Delta t c_i, t + \Delta t}$. This problem can be solved by applying a predictor-corrector method. In the predictor, we use $[K(x,t) + K(x + \Delta t c_i, t)]/2$. In the corrector we evaluate the forcing term by using the previous iterate. One corrector seemed to be sufficient; without the corrector a numerical instability occurs after an integration time of several weeks.

The forcing should respect conservation of total mass. Mass is conserved at each lattice site in the case of the 'local forcing' because

$$\sum_i c_i K(x,t) = K(x,t) \sum_i c_i = 0. \tag{5.7.13}$$

On the other hand mass is in general not conserved locally under 'non-local forcing'

$$\sum_i c_i K(x + \Delta t c_i, t) \neq 0. \tag{5.7.14}$$

Nevertheless, total mass should (and can) be conserved. Problems arise near the boundaries where some forcing terms are not compensated (in terms of mass) by contributions from other sites. This can best be illustrated by an onedimensional example (Fig. 5.7.2; all forcings at $t + \Delta t$). $K(x_1)$ does not exists because x_1 is a dry site; it is set to zero. $K(x_2)$ is not compensated. In order to compensate this term and to add some forcing at x_2, $K(x_2)$ is added at x_2 in such a way that after propagation the forcing at x_2 is comparable to that at neighboring sites (the details of the algorithm depend on the type of boundary conditions).

Fig. 5.7.2. *External forcing: a onedimensional example. The arrows indicate compensation in terms of mass.*

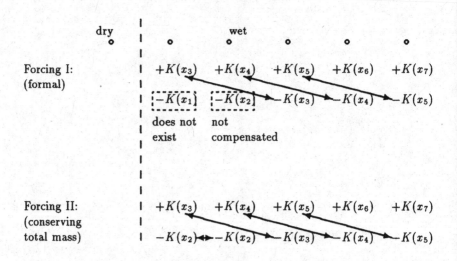

LB simulation of the linear Munk problem. The linear Munk problem was integrated in a 2000 km × 2000 km domain with central latitude at 30°N ($f_0 = 7.29 \cdot 10^{-5}$ s^{-1}, $\beta = 1.98 \cdot 10^{-11}$ s^{-1} m^{-1}). The eddy diffusivity coefficient $A = 10^4$ m^2 s^{-1} results in a width of the Munk layer $W_M = 80$ km which is almost twice the grid spacing ($\Delta x = 50$ km). The characteristic speed $U = 5 \cdot 10^{-4}$ m s^{-1} (linear regime!) is consistent with a wind stress coefficient $T_0 = 7.9 \cdot 10^{-10}$ m s^{-2}. The grid Reynolds number ($R_{e,g} = \dfrac{U \cdot \Delta x}{A} = 2.5 \cdot 10^{-3}$) is small compared to one.

In explicit numerical schemes the time step is limited by the fastest waves in the system (Courant, Friedrichs und Lewy, 1928). Rossby (planetary) waves are excited by the sudden onset of wind forcing. The free undamped waves are governed by

$$\frac{\partial}{\partial t}\nabla^2\psi + \beta\frac{\partial\psi}{\partial x} = 0. \tag{5.7.15}$$

Inserting the ansatz $\psi = \psi_0 e^{i(kx+ly-\sigma t)}$ (wavenumber $k = (k,l)$, $k, l \geq 0$) into Eq. (5.7.15) one can readily derive the dispersion relation

$$\sigma(k,l) = -\frac{\beta k}{k^2 + l^2} \tag{5.7.16}$$

which shows that Rossby waves live on the 'β effect', i.e. the fact that the rotation rate (as measured by the Coriolis parameter f) varies with latitude. The phase speed

$$v_{ph} = \frac{\sigma}{\kappa} = -\frac{\beta k}{\kappa^3} \quad \text{with} \quad \kappa = \sqrt{k^2 + l^2} \tag{5.7.17}$$

is always negativ (westward propagating phase). Its magnitude becomes maximal for $k = 2\pi/L$ and $l = 0$ and reads

$$|v_{ph}|_{max} = \frac{\beta L^2}{4\pi^2} = 2 \text{ m s}^{-1}. \tag{5.7.18}$$

Thus the time step has to below $\Delta x/|v_{ph}|_{max} \approx 2500$ s. Integration with $\Delta t = 200$ s was numerically stable over the whole simulation time (15 weeks). The global kinetic energy (Fig. 5.7.3) increases over two weeks until it reaches a quasi steady state with oscillations due to Rossby waves. The mean (over several weeks in quasi steady state) velocity shows excellent agreement with the analytical solution (Fig. 5.7.4).

LB simulation in the nonlinear regime. In order to test the LBM under nonlinear conditions, simulations were performed in a closed (no-slip boundary conditions everywhere) basin of size 4000 km × 4000 km with central latitude at 30°N. The flow can be characterized by two dimensionless numbers: the *Rossby number*

$$R_o = \frac{U}{\beta L^2} \tag{5.7.19}$$

Fig. 5.7.3. *Time series of the global kinetic energy (LBM simulation of the linear Munk problem). The oscillations in the quasi steady state are due to Rossby waves.*

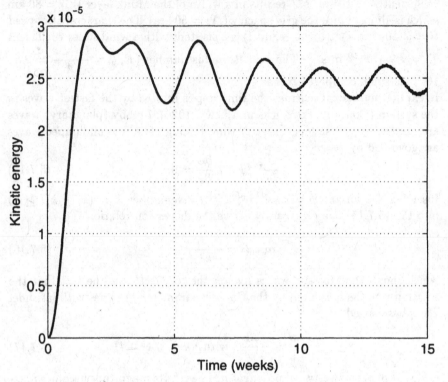

Fig. 5.7.4. *Isocontours of the velocity component u (times 10^4) show excellent agreement between the analytical solution (left) and the LBM result (right).*

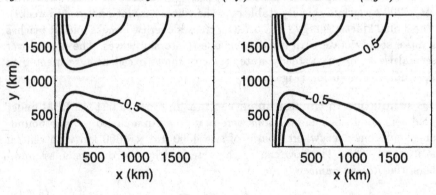

measures the ratio between the advection ($u\nabla u$) and the β term ($\beta y\hat{u}$), and the *Reynolds number*

$$R_e = \frac{U\,L}{A} \qquad (5.7.20)$$

is the ratio between the advection and the friction term ($A\nabla^2 u$). The ratio of the Rossby and Reynolds number gives the *Ekman number*

$$E_A = \frac{A}{\beta L^3} \qquad (5.7.21)$$

which measures the ratio between the friction and the β term. The Reynold number, $Re = \dfrac{U \cdot L}{A}$, was set to 80 and the Rossby number, $Ro = \dfrac{U}{\beta \cdot L^2}$, to $1.28 \cdot 10^{-3}$. $U = 0.4$ m s^{-1} and $A = 20300$ m^2 s^{-1} where calculated from Re and Ro. The grid spacing, $\Delta x = 40$ km, was chosen small than the Munk width, $W_M = 100$ km. The applied wind stress

$$T_x = -T_0 \sin^2 \frac{\pi y}{L}, \quad T_y = 0 \qquad (5.7.22)$$

with $T_0 = 8 \cdot 10^{-7}$ m s^{-2} leads to the formation of a double gyre (Fig. 5.7.5).

Fig. 5.7.5. *LB simulation of a double gyre: isocontours of the stream function $\psi(x,y)$ (times 10^{-4}).*

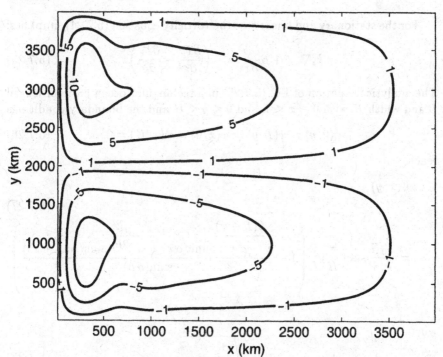

Exercise 5.7.4. (**)
A flow without any external body force can be characterized by a single dimensionless number (Reynolds number). A flow with one type of external body force requires two dimensionless numbers (Reynolds and Froude number; compare discussion of the similarity law in Section 1.3). Here we consider a flow with two types of body forces (Coriolis and wind stress). Do we need three (independent) dimensionless numbers to characterize the flow?

Exercise 5.7.5. (***)
The Navier-Stokes equation contains only molecular diffusion as dissipative process. For large scale oceanic circulation molecular diffusion does not play a role. To get rid of the vorticity imparted by the wind Stommel (1948) introduced *bottom friction* that is linear in the velocity $u = (u, v)$. The appropriate equation of motion reads

$$\frac{\partial u}{\partial t} + (u\nabla)u + f\hat{u} + \frac{1}{\rho}\nabla p + k_s u - T = 0 \qquad (5.7.23)$$

with the (bottom) friction coefficient k_s. Taking the curl of Eq. (5.7.23) one obtains the *vorticity equation*

$$\frac{\partial}{\partial t}\nabla^2\psi + J(\psi, \nabla^2\psi) + \beta\frac{\partial\psi}{\partial x} + k_s\nabla^2\psi + \left(\frac{\partial T_y}{\partial x} - \frac{\partial T_x}{\partial y}\right) = 0. \qquad (5.7.24)$$

For the stationary and linear case the vorticity equation (5.7.24) simplifies:

$$k_s\nabla^2\psi + \beta\frac{\partial\psi}{\partial x} = -\left(\frac{\partial T_y}{\partial x} - \frac{\partial T_x}{\partial y}\right). \qquad (5.7.25)$$

The analytical solution of Eq. (5.7.25) in a rectangular ocean basin of length L and width H with $0 \leq x \leq L$ and $0 \leq y \leq H$ and the boundary conditions

$$\psi(0, y) = \psi(L, y) = \psi(x, 0) = \psi(x, H) = 0. \qquad (5.7.26)$$

reads

$$\psi(x, y)$$

$$(5.7.27)$$

$$= \frac{T_0 H}{k_s \pi}\sin\left(\frac{\pi}{H}y\right)\left(1 - \frac{e^{\frac{\beta(L-x)}{2k_s}}\sinh(\alpha x) + e^{-\frac{\beta x}{2k_s}}\sinh(\alpha(L-x))}{\sinh(\alpha L)}\right)$$

where

$$\alpha = \sqrt{\frac{\beta^2}{4k_s^2} + \frac{\pi^2}{H^2}}. \qquad (5.7.28)$$

This solution describes an asymmetrical gyre with a narrow intense western boundary current and a wide slow southward drift in the eastern part of the basin.

1. Write a LBM code for the Stommel problem. 2. Compare results of the LBM code with the analytical solution of the linear Stommel model. 3. Discuss the pattern of the gyre as a function of Rossby number.

Further reading: Salmon (1999) .

5.8 A lattice Boltzmann equation for diffusion

The formulation of lattice-gas cellular automata (LGCA) for given partial differential equations is not straightforward and still requires 'some sort of magic'. Lattice Boltzmann models are much more flexible than LGCA because of the freedom in choosing equilibrium distributions with free parameters which can be set after a multi-scale expansion according to certain requirements. Here a LBM is presented for diffusion in an arbitrary number of dimensions (Wolf-Gladrow, 1995). The model is probably the simplest LBM which can be formulated. It is shown that the resulting algorithm with relaxation parameter $\omega = 1$ is identical to an explicit finite differences (EFD) formulation at its stability limit. Underrelaxation ($0 < \omega < 1$) allows stable integration beyond the stability limit of EFD. The time step of the explicit LBM integration is limited by accuracy and not by stability requirements.

The creation of LGCA for certain partial differential equations still seems to require 'some sort of magic' (compare quotation of Toffoli and Margulos, 1990, in Section 5.4). Here a simple LBM for diffusion is presented and it is shown how straightforward it is to derive such a model. In addition, the resulting algorithms are compared with an explicit finite difference (EFD) scheme.

5.8.1 Finite differences approximation

An explicit finite difference scheme for the diffusion equation

$$\boxed{\frac{\partial T}{\partial t} = \kappa \nabla^2 T} \tag{5.8.1}$$

(T is the concentration of a tracer, κ is the diffusion coefficient and ∇^2 is the Laplace operator in D dimensions in Cartesian coordinates) results from forward approximation in time and central differences in space

$$
\begin{aligned}
T^{(n+1)}_{k_1,k_2,\ldots,k_D} &= \frac{\kappa \Delta t}{(\Delta x)^2}(T^{(n)}_{k_1+1,k_2,\ldots,k_D} + T^{(n)}_{k_1-1,k_2,\ldots,k_D} + \\
&\quad \ldots + T^{(n)}_{k_1,k_2,\ldots,k_D+1} + T^{(n)}_{k_1,k_2,\ldots,k_D-1}) \\
&\quad + (1 - 2D\frac{\kappa \Delta t}{(\Delta x)^2})T^{(n)}_{k_1,k_2,\ldots,k_D}
\end{aligned}
$$

where equidistant and equal spacing in all dimensions have been assumed. The scheme is stable for

$$0 < \Delta t \leq \frac{1}{2D} \frac{(\Delta x)^2}{\kappa}$$

(see, for example, Ames, 1977; compare Subsection 5.6.2 for the case $D = 1$). At the upper stability limit the scheme becomes especially simple

$$T^{(n+1)}_{k_1,k_2,...,k_D} = \left(T^{(n)}_{k_1+1,k_2,...,k_D} + T^{(n)}_{k_1-1,k_2,...,k_D} \right.$$
$$\left. ... + T^{(n)}_{k_1,k_2,...,k_D+1} + T^{(n)}_{k_1,k_2,...,k_D-1} \right) / (2D)$$

that is T at the new time level is given by the mean over all neighbor values at the previous time level.

5.8.2 The lattice Boltzmann model for diffusion

"... it is well known that 90° rotational invariance is sufficient to yield full isotropy for *diffusive* phenomena."
Toffoli and Margulos (1990)

According to Toffoli and Margulos (1990) it is sufficient to use a square or a cubic lattice in two or three dimensions, respectively. The following model is applicable in an arbitrary number of dimensions. The grid velocities (vectors connecting neighboring grid points) are defined by

$$c_{2n-1} = (0, 0, ..., 0, 1, 0, ..., 0), \quad c_{2n} = (0, 0, ..., 0, -1, 0, ..., 0) \quad n = 1, 2, ..., D$$

where D is the dimension.

In general, the equilibrium distributions[6] $T_m^{(0)}$ depend on the conserved quantities (here only T), a number of parameters γ_k ($k = 0, 1, ..., N$) and on the direction (index m). Here, grids with only one speed are considered and the equilibrium distribution functions $T_m^{(0)}$ do not independent on m. T is given as the sum over the distribution functions T_m

$$T(\boldsymbol{x}, t) = \sum_m T_m(\boldsymbol{x}, t) = \sum_m T_m^{(0)}(\boldsymbol{x}, t) \tag{5.8.2}$$

---

[6] We will use the notation T_m instead of F_m for the distribution functions. The equilibrium distribution functions for pure diffusive problems, $T_m^{(0)}$, (diffusion of temperature or tracers) are simpler than those for flow problems, $F_m^{(0)}$.

where the summation runs over all directions $(m = 1, 2, ..., M = 2D)$.

The diffusion equation is a linear differential equation. Hence it is reasonable to use a linear ansatz for $T_m^{(0)}$

$$T_m^{(0)} = \gamma_0 + \gamma_1 T. \tag{5.8.3}$$

Inserting (5.8.3) into (5.8.2) yields

$$\boxed{T_m^{(0)} = \frac{T}{2D}} \tag{5.8.4}$$

that is, all free parameters are already fixed by the definition of the tracer concentration. The diffusion coefficient κ will result from the multi-scale expansion as described below.

5.8.3 Multi-scale expansion

The LBM is defined by the grid, the equilibrium distribution $T_m^{(0)}$ and the kinetic equation

$$T_m(\boldsymbol{x} + \boldsymbol{c}_m, t + 1) = (1 - \omega)T_m(\boldsymbol{x}, t) + \omega T_m^{(0)}(\boldsymbol{x}, t) \tag{5.8.5}$$

which states that the distribution at the new time level $(t + 1)$ at the neighboring site $(\boldsymbol{x} + \boldsymbol{c}_m)$ is a weighted sum of the distribution $T_m(\boldsymbol{x}, t)$ and the equilibrium distribution $T_m^{(0)}(\boldsymbol{x}, t)$. Models with parameter ω go under various names: 'enhanced collision' (Higuera et al., 1989), BGK (named after Bhatnagar, Gross and Krook, 1954; compare, for example, Qian et al., 1992), STRA ('single time relaxation approximation', Chen et al., 1991) or SOR ('successive over-relaxation', Qian et al., 1992). The LBM is stable for $0 < \omega < 2$. Now the macroscopic equations will be derived by a multi-scale analysis (compare Frisch et al., 1987, for an analogous procedure for LGCA). The distribution functions are expanded up to linear terms in the small expansion parameter ϵ

$$T_m = T_m^{(0)} + \epsilon T_m^{(1)} + \mathcal{O}(\epsilon^2).$$

From the kinetic equation (5.8.5) one can calculate an approximation of $T_m^{(1)}$

$$
\begin{aligned}
T_m(\boldsymbol{x} + \boldsymbol{c}_m, t + 1) &= T_m(\boldsymbol{x}, t) + \partial_{x_\alpha} c_{m\alpha} T_m + \partial_t T_m + \mathcal{O}(\epsilon^2) \\
&= (1 - \omega) \underbrace{T_m(\boldsymbol{x}, t)} + \omega T_m^{(0)}(\boldsymbol{x}, t) \\
&= T_m^{(0)} + \epsilon T_m^{(1)} + \mathcal{O}(\epsilon^2)
\end{aligned}
$$

$$\longrightarrow$$

$$\epsilon T_m^{(1)} = -\frac{1}{\omega} \partial_{x_\alpha} c_{m\alpha} T_m - \frac{1}{\omega} \partial_t T_m + \mathcal{O}(\epsilon^2).$$

Diffusion is a slow process on large spatial scales which suggests the following scaling (same as for the derivation of the Navier-Stokes equations in Frisch et al., 1987)

$$\partial_t \;\; \rightarrow \;\; \epsilon^2 \partial_t^{(2)}$$
$$\partial_{x_\alpha} \;\; \rightarrow \;\; \epsilon \partial_{x_\alpha}^{(1)}.$$

The components of the grid velocities obey the following equations

$$\sum_m c_m = 0$$

$$\sum_m c_{m\alpha} c_{m\beta} = 2\delta_{\alpha\beta}$$

and therefore

$$\sum_m c_m T_m^{(0)} = \frac{T}{2D} \sum_m c_m = 0.$$

Inserting the expansion and the scalings into the conservation relation for tracer concentration, one obtains up to second order in ϵ

$$
\begin{aligned}
0 &= \sum_m [T_m(\boldsymbol{x} + \boldsymbol{c}_m, t+1) - T_m(\boldsymbol{x}, t)] \\
&= \sum_m [T_m(\boldsymbol{x}, t) + \underbrace{\epsilon^2 \partial_t^{(2)} T_m}_{\rightarrow \partial_t T} + \epsilon \partial_{x_\alpha}^{(1)} c_{m\alpha} T_m + \frac{1}{2} \epsilon^2 \partial_{x_\alpha}^{(1)} \partial_{x_\beta}^{(1)} c_{m\alpha} c_{m\beta} T_m^{(0)} \\
&\quad - T_m(\boldsymbol{x}, t) + \mathcal{O}(\epsilon^3)]
\end{aligned}
$$

and

$$
\begin{aligned}
\sum_m \epsilon \partial_{x_\alpha}^{(1)} c_{m\alpha} T_m &= \epsilon \partial_{x_\alpha}^{(1)} \underbrace{\sum_m c_{m\alpha} T_m^{(0)}}_{= 0} + \sum_m \epsilon^2 \partial_{x_\alpha}^{(1)} c_{m\alpha} T_m^{(1)} + \mathcal{O}(\epsilon^3) \\
&= -\frac{1}{\omega} \sum_m \epsilon^2 \partial_{x_\alpha}^{(1)} \partial_{x_\beta}^{(1)} c_{m\alpha} c_{m\beta} T_m^{(0)} + \mathcal{O}(\epsilon^3) \qquad (5.8.6) \\
&= -\frac{1}{\omega} \frac{1}{D} \underbrace{\epsilon^2 \delta_{\alpha\beta} \partial_{x_\alpha}^{(1)} \partial_{x_\beta}^{(1)} T}_{\rightarrow \nabla^2 T} + \mathcal{O}(\epsilon^3)
\end{aligned}
$$

$$\sum_m \frac{1}{2} \epsilon^2 \partial_{x_\alpha}^{(1)} \partial_{x_\beta}^{(1)} c_{m\alpha} c_{m\beta} T_m^{(0)} = \frac{1}{2D} \epsilon^2 \delta_{\alpha\beta} \partial_{x_\alpha}^{(1)} \partial_{x_\beta}^{(1)} T$$

and finally

$$\frac{\partial T}{\partial t} = \kappa \nabla^2 T$$

with

$$\kappa = \left(\frac{1}{\omega} - \frac{1}{2}\right)\frac{1}{D}.$$

(5.8.7)

5.8.4 The special case $\omega = 1$

For $\omega = 1$ the kinetic equation (5.8.5) reduces to

$$T_m(x + c_m, t+1) = T_m^{(0)}(x, t)$$

and the diffusion coefficient is $\kappa = \dfrac{1}{2D}$. This LBM is identical to the finite difference scheme at the stability limit. The right hand side of the kinetic equation is just the mean value of the nearest neighboring sites and the diffusion coefficient is the maximal value allowed by the stability condition. For the LBM the diffusion coefficient is expressed in the units $\Delta t = \Delta x = 1$; the diffusion coefficient at the stability limit of the EFD reads

$$\kappa = \frac{1}{2D}\frac{(\Delta x)^2}{\Delta t} = \frac{1}{2D}.$$

This scheme requires only two arrays in memory: the tracer concentrations at two time levels.

5.8.5 The general case

In the general case one has to store $M = 2D$ distributions in addition to the tracer concentrations at two time levels. What do we gain from this extra cost? In the range $0 < \omega < 1$ (underrelaxation) the diffusion coefficient κ is larger than the value at the stability limit of the EFD scheme while we still keep $\Delta t = \Delta x = 1$. In contrast to EFD, the LBM is stable in this parameter range.

5.8.6 Numerical experiments

To test the predictions of the LBM outlined above the one-dimensional diffusion equation was integrated. As initial conditions, values of an analytical solution were used, namely

$$T(x, t_i) = \frac{1}{2\sqrt{\pi \kappa t_i}}\exp\left[-\frac{x^2}{4\kappa t_i}\right].$$

The integration starts at $t_i = 15/\kappa(\omega)$ and ends at $t_f = 75/\kappa(\omega)$, thus the time interval depends on $\kappa(\omega)$, but in each case the integration starts with the same numerical values and ends after the maximum decreases from ≈ 0.073 to ≈ 0.033. Fig. 5.8.1 shows the results of such an integration for $\omega = 0.3$ together with the analytical solution and the initial values. By appropriate choice of ω, one can keep $\Delta t = 1$ for 'arbitrarily' large diffusion coefficients: the scheme is stable but the numerical error increases with increasing diffusion coefficient (compare Fig. 5.8.2). Thus, we have an explicit scheme (BGK-LBM) where the length of the time step is no longer limited by stability requirements. The large error at small values of ω stems from the fact that explicit approximations of parabolic equations act like a hyperbolic system with two real finite difference characteristics instead of only a single real characteristic of the continuous system (Ames, 1977).

5.8.7 Summary and conclusion

A very simple LBM for diffusion in an arbitrary number of dimensions is proposed. For $\omega = 1$ the resulting algorithm is identical to an explicit finite difference scheme at its stability limit. Thus the LBM scheme is not only stable, but automatically picks the maximal allowed diffusion coefficient κ to ensure stability of the EFD scheme.

For LGCA the transport coefficients depend on the collision rules which are never optimal in the sense that they yield only a certain approximation of the (continuous) local equilibrium functions (compare the various FHP models with and without rest particles in Frisch et al., 1987, or the various collision rules proposed for FCHC by Hénon, 1987, Rem and Somers, 1989, and van Coevorden et al., 1994) whereas for LBM, the collisions (which do not show up explicitly) can create local equilibrium at each time step. By reducing the number of collision in LGCA one obtains models with higher diffusion coefficients while stability is assured. This can be regarded as a kind of underrelaxation.

In the BGK-LBM the diffusion coefficient κ is an adjustable parameter. Of special interest is the parameter range $0 < \omega < 1$. The use of information contained in the nonequilibrium distribution functions allows explicit stable integration beyond the stability limit of the EFD scheme. Thus, the time step is limited by accuracy and not by stability requirements.

Fig. 5.8.1. *Integration of the diffusion equation in one dimension by the BGK-LBM with $\omega = 0.3$. The integration starts at time $t_i = 15/\kappa(\omega)$ with initial values $T(x, t_i) = \dfrac{1}{2\sqrt{\pi \kappa t_i}} \exp\left[-\dfrac{x^2}{4\kappa t_i}\right]$ (dotted line) and ends at $t_f = 75/\kappa(\omega)$. The figure shows the numerical results (broken line) together with the analytical solution (solid line).*

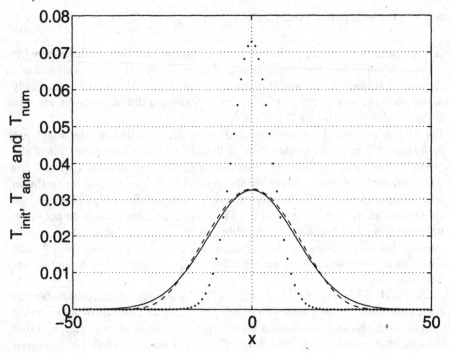

Fig. 5.8.2. *Integrations of the diffusion equation in one dimension by the BGK-LBM. The integration starts at time $t_i = 15/\kappa(\omega)$ with initial values $T(x, t_i) = \dfrac{1}{2\sqrt{\pi \kappa t_i}} \exp\left[-\dfrac{x^2}{4\kappa t_i} \right]$ and ends at $t_j = 75/\kappa(\omega)$. The plot shows the logarithm of the maximum error ($max\{|T_{numerical\ solution} - T_{analytical\ solution}|\}$) at the end of the integrations as a function of ω. The error increases at small values of ω (large values of the diffusion coefficients).*

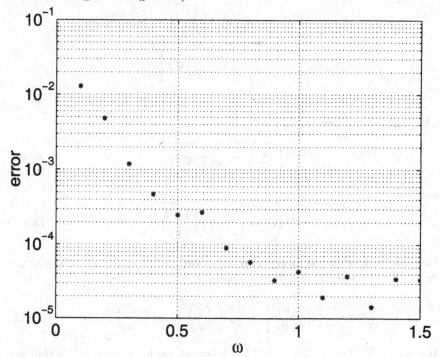

5.8.8 Diffusion equation with a diffusion coefficient depending on concentration

By a small modification the lattice Boltzmann model proposed above can be generalized to a model for a diffusion equation with a diffusion coefficient depending on concentration: The constant ω will be replaced by a function $\omega(T)$. The multi-scale analysis proceeds as before except for Eq. (5.8.6) where $1/\omega$ and the spatial derivative must not be exchanged:

$$\sum_m \epsilon \partial_{x_\alpha}^{(1)} c_{m\alpha} T_m = \epsilon \partial_{x_\alpha}^{(1)} \underbrace{\sum_m c_{m\alpha} T_m^{(0)}}_{= 0} + \sum_m \epsilon^2 \partial_{x_\alpha}^{(1)} c_{m\alpha} T_m^{(1)} + \mathcal{O}(\epsilon^3)$$

$$= -\sum_m \epsilon^2 \partial_{x_\alpha}^{(1)} \frac{1}{\omega(T)} \partial_{x_\beta}^{(1)} c_{m\alpha} c_{m\beta} T_m^{(0)} + \mathcal{O}(\epsilon^3)$$

$$= -\frac{1}{D} \epsilon^2 \delta_{\alpha\beta} \partial_{x_\alpha}^{(1)} \frac{1}{\omega(T)} \partial_{x_\beta}^{(1)} T + \mathcal{O}(\epsilon^3).$$

$$\rightarrow \left[\nabla \frac{1}{\omega(T)} \nabla T \right]$$

Thus the macroscopic equation reads

$$\boxed{\frac{\partial T}{\partial t} = \nabla \left[\kappa(T) \nabla T \right]} \tag{5.8.8}$$

with

$$\boxed{\kappa(T) = \left[\frac{1}{\omega(T)} - \frac{1}{2} \right] \frac{1}{D}.} \tag{5.8.9}$$

Comparison with an analytical solution. A few analytical solutions are known for the *nonlinear diffusion equation*

$$\frac{\partial T}{\partial t} = \frac{\partial}{\partial x} \left[\kappa(T) \frac{\partial T}{\partial x} \right] = \frac{\partial \kappa}{\partial T} \left(\frac{\partial T}{\partial x} \right)^2 + \kappa(T) \frac{\partial^2 T}{\partial x^2}. \tag{5.8.10}$$

For $\kappa(T) = T$ (Logan, 1994, p. 143; a misprint has been corrected) a solution reads

$$T(x,t) = \frac{1}{6t} \left(A^2 t^{2/3} - x^2 \right) \quad \text{for} \quad |x| < At^{1/3} \tag{5.8.11}$$

and $T(x,t) = 0$ otherwise. Fig. 5.8.3 shows the comparison between the numerical solution by the lattice Boltzmann model and the analytical solution at $t = 210$ for initial conditions according to eq. (5.8.11) with $A = 5$, $t = 10$ inside the interval $|x| < At^{1/3}$ and $T = 0$ otherwise. The agreement is very good.

Fig. 5.8.3. *Integration of the nonlinear diffusion equation (5.8.10) by the lattice Boltzmann model. The initial distribution is marked by circles. The numerical solution at t = 210 (solid line) is indistinguishable from the analytical solution (dashed-dotted line; not visible). The broken line shows the difference between numerical and analytical solution multiplied by 100.*

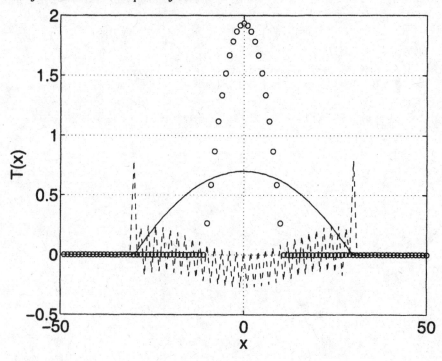

5.8.9 Further reading

Elton et al. (1990) proposed a LBM for a nonlinear diffusion equation in 2D. The collision operator is based on variants of the HPP model (extended by one- and two-particle collisions which conserve mass but not momentum). The diffusion coefficients are functions of the tracer density.

The LBM for reaction-diffusion systems by Dawson et al. (1993) lives on the FHP lattice (D2Q7) and applies the BGK approximation of the collision operator.

Qian and Orszag (1995) developed a LBM for reaction-diffusion equations. Their diffusion model is identical to that proposed by Wolf-Gladrow (1995).

Further reading: Sun (1998), van der Sman and Ernst (1999).

Finite differences: Teixeira (1999).

5.9 Lattice Boltzmann model: What else?

– Books: Rothman and Zaleski (1997).

– Review articles: Benzi, Succi and Vergassola (1992), Rothman and Zaleski (1994), Qian, Succi and Orszag (1995), Biggs and Humby (1998), Chen and Doolen (1998).

– Curvilinear coordinates, finite volumes, irregular grids, mesh refinement: Filippova and Hänel (1998a,b), He and Doolen (1997a,b), He (1997), Karlin and Succi (1998), Mei and Shyy (1998), Peng et al. (1998, 1999), Renwei and Wei (1998), Tölke et al. (1998, 1999), Karlin et al. (1999), Xi et al. (1999a,b).

– Flow past obstacles: Bernsdorf et al. (1998).

– Flow through porous media: Cancelliere et al. (1990), Aharonov and Rothman (1993), Gunstensen and Rothman (1993), Sahimi (1993), Buckles et al. (1994), Heijs and Lowe (1995), Martys and Chen (1996), Shan and Doolen (1996), Grubert (1997), Spaid and Phelan (1997; 1998), Angelopoulos et al. (1998), Bosl et al. (1998), Coveney et al. (1998), Dardis and McCloskey (1998a,b), Freed (1998), Kim et al. (1998), Koch et al. (1998), Koponen et al. (1998a,b), Maier et al. (1998a,b), Noble and Torczynski (1998), Waite et al. (1998; 1999), Berest et al. (1999), Bernsdorf et al. (1999), Inamuro et al. (1999), Verberg and Ladd (1999).

– Granular flow: Herrmann (1995), Tan et al. (1995), Herrmann et al. (1996).

– Multiphase flows: Gunstensen et al. (1991), Holme and Rothman (1992), Flekkøy (1993), Shan and Chen (1993, 1994), d'Ortona et al. (1995), Flekkøy et al. (1995), Orlandini et al. (1995), Flekkøy et al. (1996b), Halliday (1996), Halliday et al. (1996), Sofonea (1996), Swift et al. (1996), Gonnella et al. (1997; 1998; 1999), Hou et al. (1997), Kato et al. (1997), Wagner and Yeomans (1997; 1998; 1999), Angelopoulos et al. (1998), Chen et al. (1998a), Halliday et al. (1998), Holdych et al. (1998), Lamura et al. (1998; 1999), Masselot and Chopard (1998a), Theissen et al. (1998), Wagner (1998b), Yu and Zhao (1999).

– Magnetohydrodynamics (MHD): Chen et al. (1991), Succi et al. (1991), Martinez et al. (1994), Sofonea (1994).

– Compressible flows, Burgers equation: Alexander et al. (1992).

– Flow past fractal obstacles: Adrover and Giona (1997).

– Turbulence and large eddy simulation (LES): Frisch (1991), Benzi et al. (1996), Amati et al. (1996), Fogaccia et al. (1996), Hayot and Wagner (1996), Amati et al. (1997a,b), Succi (1998), Teixeira (1998).

– Flow in dynamical geometry (blood flow): Fang et al. (1998), Krafczyk et al. (1998a).

- Glacier flow: Bahr and Rundle (1995).

- Rayleigh-Bénard convection: Bartoloni et al. (1993), Massaioli et al. (1993), Benzi et al. (1994), Pavlo et al. (1998a) Vahala et al. (1998a).

- Rayleigh-Taylor instability: Nie et al. (1998), He et al. (1999a,b).

- Korteweg-de Vries equation: Yan (1999).

- Droplets: Schelkle et al. (1999), Xi and Duncan (1999).

- Crystal growth: Miller and Böttcher (1998).

- Wave propagation: Chopard and Luthi (1999).

- Maxwell's equations: Simons et al. (1999).

- Decoupling of spatial grid from the velocity lattice: Cao et al. (1997), Pavlo et al. (1998a).

- H-theorem for LBM: Karlin and Succi (1998), Karlin et al. (1998), Wagner (1998a), Karlin et al. (1999).

- Quantum mechanics: Succi and Benzi (1993), Succi (1996), Meyer (1997a,b; 1998).

- Related methods: Junk (1999).

- Further reading: Lavallée et al. (1989), Kingdon et al. (1992), McNamara and Alder (1992, 1993), Nannelli and Succi (1992), Qian (1993), Qian and Orszag (1993), Ladd (1993, 1994a,b), Chen et al. (1994), Punzo et al. (1994), Succi and Nannelli (1994), Elton et al. (1995), Hou et al. (1995), McNamara et al. (1995), Miller (1995), Ohashi et al. (1995), Reider and Sterling (1995), Succi et al. (1995), Wagner and Hayot (1995), Zou et al. (1995a,b), Flekkøy et al. (1996a), Elton (1996), (Burgers eq.) He et al. (1996), Kaandorp et al. (1996), Lin et al. (1996), Orszag et al. (1996), Rakotomalala et al. (1996), Sterling and Chen (1996), Chen and Ohashi (1997), Filippova and Hänel (1997, 1998c), Giraud and d'Humières (1997), Luo (1997a,b), Maier and Bernard (1997), Qi (1997; 1999), Qian (1997), Qian and Chen (1997), Stockman et al. (1997; 1998), Succi et al. (1997), van der Sman (1997), Warren (1997), Ahlrichs and Dunweg (1998), Aidun et al. (1998), Buick and Greated (1998), Buick et al. (1998), Chen (1998), Chen and Ohashi (1998), Chen et al. (1998b), Chenghai (1998), De Fabritiis et al. (1998), Giraud et al. (1998), He and Zhao (1998), Kandhai et al. (1998a,b; 1999), Krasheninnikov and Catto (1998), Luo (1998), Succi and Vergari (1998), Takada and Tsutahara (1998), Ujita et al. (1998), Xu and Luo (1998), Yan (1998), Yan et al. (1998; 1999), Kendon et al. (1999), Klar (1999).

5.10 Summary and outlook

Because LGCA and LBM are still in rapid development it is not possible to give an actual and complete picture of the whole field. Instead I have tried to introduce the basic models (HPP, FHP, FCHC, D2Q9) plus some personal favorites (like PI) together with methods from statistical mechanics (Chapman-Enskog expansion, BGK approximation, maximum entropy principle) which are necessary for the theory of LGCA or LBM but which are usually not part of the curriculum for students of physics or mathematics. Knowledge of these special methods is usually taken for granted in articles and even in reviews.

I am still fascinated by these new methods, and I guess, I am not alone. Although the emergence of new numerical methods is often driven by practical requirements by engineers and natural scientists, (applied) mathematicians usually take the lead in the development of the schemes. However, many researchers working on LGCA and LBM would call themselves most probably physicists. Why are these kind of schemes so attractive to them? At least three reason come to my mind. Firstly, both approaches are based on conservation laws. Physicists feel at home with conservation laws. The continuity equation and the Navier-Stokes equation express the conservation of mass and momentum. Certain conserved quantities are related (Noether theorem) to symmetry groups which allow us to derive equations of the fundamental field theories by the gauge principle. Secondly, LGCA and LBM require more (physical) theory than many other numerical methods. Whereas you can teach to a beginner how to create and apply the simplest schemes of finite differences or spectral methods to partial differential in a few hours, this seems not possible for LGCA, LBM, and finite elements. And the theory required stems from statistical mechanics (Chapman-Enskog expansion, maximum entropy principle) whereas for finite elements, for example, weak convergence in Sobolev spaces is more the backyard of mathematicians. Last but not least, it is fascinating to see the role of symmetry. The symmetry of the underlying lattice is still important on the macroscopic level. It took more than 10 years to discover that it is sufficient to replace a lattice with fourfold (HPP) by one with hexagonal (FHP) symmetry in order to obtain the correct form of the nonlinear advection term of the Navier-Stokes equation. The same symmetry requirement even led us into four dimensions (FCHC). Chiral symmetry has to be established by random processes (FHP). Spurious invariants have to be detected (still not solved in general) and then to be destroyed (for example: three-particle collisions in FHP) or not to be initialized or generated (Zanetti invariants). And finally consequences of missing symmetry has to be scaled away (symptomatic treatment of the violation of Galilean invariance).

As a final remark let me ask the question: What will be the future of LGCA and LBM? Will these schemes have outcompeted the traditional methods (finite differences, finite volumes, finite elements, spectral, semi-Lagrange, ...) in say five years from now? Or will we throw away these new schemes because they cannot compete with the fastest of the other schemes? The application of a wide variety of methods at the same time already indicates that currently (forever?) there does not exist a best scheme for all problems in the numerical solution of partial differential equations. Although finite elements are well established since decades (the first edition of the monograph by Zienkiewicz appeared already in 1967) the Modular Ocean Model (the work horse of physical oceanography) still applies finite volumes (Bryan, 1969; Semtner, 1997). And the traditional methods are still under development (the same is true for languages such as FORTRAN). Rood (1987) lists about 100 different numerical advection schemes and several new have been proposed since. LGCA and LB models have disadvantages (how to construct models for given differential equations, noise of LGCA, stability of LBM, large memory requirement of LBM) which prohibit the general replacement of well-established traditional methods.

On the other hand, LGCA and LB models have several advantages like flexibility with respect to domain geometry, locality of the collisions (well adapted to massively parallel computer) and low complexity of the code. Because LB models are based on conservation principles they are especially suited for long-time integrations (for instance: climate models). It has been shown already that LGCA and LB models can compete with traditional schemes in certain classes of problems. Therefore, it is almost certain that both methods will find their niches.

6. Appendix

6.1 Boolean algebra

George Boole (1815 - 1864), English mathematician. Boolean algebra is a mathematical structure $\mathcal{B} = (B, \cup, \cap, ^-)$ consisting of a set B and two binary operations called *union* ($\cup$) and *intersection* ($\cap$) and one unitary operation called *complementation* ($^-$). The system which originally inspired this collection of laws is the algebra of sets with the familiar operations of set union and intersection.

The following relations[1] hold true:

1. Closure:

   (i) The union of two elements in B yields a unique element in B.
      $a, b \in B \rightarrow (a \cup b) \in B$

   (ii) The intersection of two elements in B yields a unique element in B.
      $a, b \in B \rightarrow (a \cap b) \in B$

2. Commutativity: $\forall a, b \in B$:

   (i) $a \cup b = b \cup a$

   (ii) $a \cap b = b \cap a$

3. Associativity: $\forall a, b, c \in B$:

   (i) $(a \cup b) \cup c = a \cup (b \cup c)$

   (ii) $(a \cap b) \cap c = a \cap (b \cap c)$

4. Distributivity: $\forall a, b, c \in B$:

   (i) $a \cap (b \cup c) = (a \cap b) \cup (a \cap c)$

   (ii) $a \cup (b \cap c) = (a \cup b) \cap (a \cup c)$

5. The idempotent laws: $\forall a \in B$

   (i) $a \cup a = a$

   (ii) $a \cap a = a$

6. Identity elements 0 and 1:

   (i) In B there is the unique element 0 with the following properties:
      $a \cup 0 = a$ and $a \cap 0 = 0 \; \forall a \in B$.
      0 is the identity element with respect to union.

   (ii) In B there is the unique element 1 with the following properties:
      $a \cup 1 = 1$ and $a \cap 1 = a \; \forall a \in B$.
      1 is the identity element with respect to intersection.

   Remark: If you consider the ordinary union and intersection of sets the 0-element corresponds to the empty set and the 1-element corresponds to the whole set.

---

[1] We do not speak of axioms because some of relations are redundant.

7. Complementation: $\forall\, a \in B$ there exists a unique element $\bar{a}$ such that
   (i) $a \cup \bar{a} = 1$
   (ii) $a \cap \bar{a} = 0$.
   $\bar{a}$ is called the *complement* of a. The elements $a, b \in B$ obey the DeMorgan laws:
   (iii) $\overline{a \cup b} = \bar{a} \cap \bar{b}$
   (iv) $\overline{a \cap b} = \bar{a} \cup \bar{b}$
   Law of involution: $\forall\, a \in B$
   (v) $\overline{(\bar{a})} = a$

Principle of duality: The substitution

$$(\cup, \cap, 0, 1) \quad \rightarrow \quad (\cap, \cup, 1, 0)$$

transforms true expressions of the Boolean algebra into true expressions. Proof: All laws listed above obey the principle of duality. Law 7(v) is selfdual.

Exercise 6.1.1. (**)
Show that the structure

$$B_{tf} = (B = \{true, false\}, AND, (\text{inclusive})OR, NOT)$$

defines a Boolean algebra. Show that XOR can be expressed in terms of AND, OR and NOT. Why is

$$\mathcal{N} = (B = \{true, false\}, AND, XOR, NOT)(XOR = \text{ exclusive or})$$

not a Boolean algebra?

Exercise 6.1.2. (*)
Consider the set $B = \{0, 1 \in \mathbb{N}\}$ and the operations addition modulo 2 $(+)$ and multiplication $(\cdot)$. Show: this structure obeys the laws of Boolean algebra if complementation is appropriately defined (how?).

Exercise 6.1.3. (**)
Prove that for all elements a and b in the set B of a Boolean algebra $(B, \cup, \cap, ^-)$:

$$(a \cap b) \cup (a \cap \bar{b}) = a$$

Exercise 6.1.4. (*)
Consider the set of integers and the operations addition $(+)$ and multiplication $(\cdot)$. Show that this structure is not a Boolean algebra.

6.2 FHP: After some algebra one finds ...

In this appendix the Lagrange multipliers of the equilibrium distributions for the FHP-I and the HPP model will be calculated (compare Subsection 3.2.5). For vanishing flow velocity the occupation numbers equal each other

$$\boldsymbol{u} = 0 \quad \Longrightarrow \quad N_i = \frac{\rho}{b} = d \tag{6.2.1}$$

(b number of cells per node, $b = 6$ for FHP-I, $b = 4$ for HPP; d density per cell) and therefore

$$N_i(\rho, 0) = \frac{1}{1 + \exp[h(\rho, 0)]} = d \tag{6.2.2}$$

and

$$q(\rho, 0) = 0. \tag{6.2.3}$$

Because of invariance of the occupation numbers under the parity transform

$$\boldsymbol{u} \quad \rightarrow \quad -\boldsymbol{u}, \quad \boldsymbol{c}_i \quad \rightarrow \quad -\boldsymbol{c}_i \tag{6.2.4}$$

it follows that

$$h(\rho, -\boldsymbol{u}) = h(\rho, \boldsymbol{u}), \tag{6.2.5}$$

and

$$q(\rho, -\boldsymbol{u}) = -q(\rho, \boldsymbol{u}). \tag{6.2.6}$$

The Lagrange multipliers h und q will be expanded up to second order in $\boldsymbol{u}$:

$$h(\rho, \boldsymbol{u}) = h_0(\rho) + h_2(\rho)\boldsymbol{u}^2 + \mathcal{O}(\boldsymbol{u}^4) \tag{6.2.7}$$
$$q(\rho, \boldsymbol{u}) = q_1(\rho)\boldsymbol{u} + \mathcal{O}(\boldsymbol{u}^3). \tag{6.2.8}$$

All other low order terms vanish because of the parity constraints. It is remarkable that h_2 and q_1 are scalars instead of tensors of rank 2. This fact is a consequence of the isotropy of lattice tensors (compare Section 3.3) of rank 2.

The expansions (6.2.7) and (6.2.8) will be inserted into the Fermi-Dirac distribution (3.2.28). Then the distribution is expanded in a Taylor series with respect to $\boldsymbol{u}$ at $\boldsymbol{u} = 0$ up to second order.

$$N_i(\boldsymbol{u}) = N_i(\boldsymbol{u} = 0) + \frac{\partial N_i}{\partial u} \cdot u + \frac{\partial N_i}{\partial v} \cdot v$$
$$+ \frac{1}{2}\frac{\partial^2 N_i}{\partial u^2} \cdot u^2 + \frac{\partial^2 N_i}{\partial u \partial v} \cdot u \cdot v + \frac{1}{2}\frac{\partial^2 N_1}{\partial v^2} \cdot v^2 + \mathcal{O}(\boldsymbol{u}^3),$$

$$N_i(\boldsymbol{u}) = \frac{1}{1 + \exp[x(\boldsymbol{u})]},$$

$$x(\boldsymbol{u}) = h_0 + h_2\boldsymbol{u}^2 + q_1\boldsymbol{u}c_i,$$

$$x(0) = h_0,$$

$$N_i(h_0) = \frac{1}{(1 + \exp[h_0])} = d,$$

$\longrightarrow$

$$\exp[h_0] = \frac{1-d}{d},$$

$$h_0 = \ln\frac{1-d}{d}$$

$$\frac{\partial N_i}{\partial u_\alpha} = \frac{\partial N_i}{\partial x} \cdot \frac{\partial x}{\partial u_\alpha} \underset{\boldsymbol{u}=0}{\longrightarrow} d(d-1)q_1c_{i\alpha},$$

$$\frac{\partial N_i}{\partial x} = -\frac{\exp[x]}{(1+\exp[x])^2} \underset{\boldsymbol{u}=0}{\longrightarrow} \frac{d-1}{d}d^2 = d(d-1)$$

$$\frac{\partial x}{\partial u_\alpha} = 2h_2u_\alpha + q_1c_{i\alpha} \longrightarrow q_1c_{i\alpha},$$

$$\frac{\partial^2 N_i}{\partial u_\alpha^2} = \frac{\partial}{\partial u_\alpha}\left[\frac{\partial N_i}{\partial x} \cdot \frac{\partial x}{\partial u_\alpha}\right]$$

$$= \frac{\partial^2 N_i}{\partial x \partial u_\alpha} \cdot \frac{\partial x}{\partial u_\alpha} + \frac{\partial N_i}{\partial x} \cdot \frac{\partial^2 x}{\partial u_\alpha^2}$$

$$= \frac{\partial^2 N_i}{\partial x^2} \cdot \left(\frac{\partial x}{\partial u_\alpha}\right)^2 + \frac{\partial N_i}{\partial x}\frac{\partial^2 x}{\partial u_\alpha^2}$$

$$\longrightarrow d(d-1)(2d-1)q_1^2c_{i\alpha}^2 + d(d-1)2h_2,$$

$$\frac{\partial^2 N_i}{\partial x^2} = -\frac{\exp[x](1+\exp[x])^2 - \exp[x] \cdot 2(1+\exp[x])\exp[x]}{(1+\exp[x])^4}$$

$$= \frac{\exp[x](\exp[x]-1)}{(1+\exp[x])^3}$$

$$\longrightarrow \frac{1-d}{d}(\frac{1-d}{d}-1)d^3 = d(d-1)(2d-1),$$

$$\frac{\partial^2 x}{\partial u_\alpha^2} = 2h_2.$$

For $\alpha \neq \beta$:

$$\frac{\partial^2 N_i}{\partial u_\alpha \partial u_\beta} = \frac{\partial}{\partial u_\beta}\left[\frac{\partial N_i}{\partial x} \cdot \frac{\partial x}{\partial u_\alpha}\right]$$

$$= \frac{\partial^2 N_i}{\partial x \partial u_\beta} \cdot \frac{\partial x}{\partial u_\alpha} + \frac{\partial N_i}{\partial x} \cdot \frac{\partial^2 x}{\partial u_\alpha \partial u_\beta}$$

$$= \frac{\partial^2 N_i}{\partial x^2} \cdot \frac{\partial x}{\partial u_\alpha} \cdot \frac{\partial x}{\partial u_\beta} + \frac{\partial N_i}{\partial x} \cdot \frac{\partial^2 x}{\partial u_\alpha \partial u_\beta}$$

$$\to d(d-1)(2d-1)q_1^2 c_{i\alpha} c_{i\beta},$$

$$\frac{\partial^2 x}{\partial u_\alpha \partial u_\beta} = 0,$$

$$N_i(\boldsymbol{u}) = d + d(d-1)q_1 \boldsymbol{c}_i \cdot \boldsymbol{u}$$
$$+ \frac{1}{2}d(d-1)(2d-1)q_1^2 c_{i\alpha}^2 u_\alpha^2 + d(d-1)h_2 u^2. \quad (6.2.9)$$

At this point the coefficients h_2 and q_1 are not known yet. The expression for N_i, however, looks much simpler now: a polynomial instead of a rational function with an exponential function in the denominator. Next insert the N_i according to eq. (6.2.9) into the definitions of mass and momentum density and use the moments relations (3.2.3 - 3.2.5):

$$\rho = \sum_i N_i$$

$$= \underbrace{\sum_i d}_{= \rho} + \underbrace{\sum_i d(d-1)q_1 \boldsymbol{c}_i \cdot \boldsymbol{u}}_{= 0}$$

$$+ \underbrace{\sum_i \frac{1}{2}d(d-1)(2d-1)q_1^2 c_{i\alpha}^2 u_\alpha^2}_{= \frac{1}{2}3d(d-1)(2d-1)q_1^2 u^2}$$

$$+ \underbrace{\sum_i d(d-1)h_2 u^2}_{= 6d(d-1)h_2 u^2}.$$

From this a relation between h_2 and q_1 follows:

$$h_2 = \frac{1}{4}(1-2d)q_1^2,$$

$$\rho \boldsymbol{u} = \sum_i N_i \boldsymbol{c}_i$$

$$= \underbrace{\sum_i d\boldsymbol{c}_i}_{= 0} + \underbrace{\sum_i d(d-1)q_1(\boldsymbol{c}_i \cdot \boldsymbol{u})\boldsymbol{c}_i}_{= 3d(d-1)q_1 \boldsymbol{u}}$$

$$+ \sum_i \underbrace{\frac{1}{2}d(d-1)(2d-1)q_1^2 c_{i\alpha}^2 u_\alpha^2 c_i}_{=\,0}$$

$$+ \sum_i \underbrace{d(d-1)h_2 \boldsymbol{u}^2 c_i,}_{=\,0}$$

$$\begin{aligned}
q_1 &= \frac{2}{d-1}, \\
h_2 &= \frac{1-2d}{(d-1)^2}, \\
N_i(\boldsymbol{u}) &= d + 2d\,c_i \cdot \boldsymbol{u} + 2d\frac{1-2d}{1-d}c_{i\alpha}^2 u_\alpha^2 - d\frac{1-2d}{1-d}\boldsymbol{u}^2 \\
&= \frac{\rho}{6} + \frac{\rho}{3}c_i \cdot \boldsymbol{u} + \rho G(\rho)Q_{i\alpha\beta}u_\alpha u_\beta,
\end{aligned}$$

with

$$G(\rho) = \frac{1}{3}\frac{6-2\rho}{6-\rho} \quad \text{and} \quad Q_{i\alpha\beta} = c_{i\alpha}c_{i\beta} - \frac{1}{2}\delta_{\alpha\beta}.$$

6.3 Coding of the collision operator of FHP-II and FHP-III in C

```
/* collision with 1 rest particle (FHP-II) 17.5.89 dwg */

for(ix=0; ix < IXM; ix++) {
for(iy=0; iy < IYM; iy++) {

/* i's -> register a,...,f  */

    a = i1[ix][iy];
    b = i2[ix][iy];
    c = i3[ix][iy];
    d = i4[ix][iy];
    e = i5[ix][iy];
    f = i6[ix][iy];
    r = rest[ix][iy];

/* three body collision <-> 0,1 (bits) alternating
   <-> triple = 1 */

    triple = (a^b)&(b^c)&(c^d)&(d^e)&(e^f);

/* two-body collision
   <-> particles in cells a (b,c) and d (e,f)
       no particles in other cells
   <-> db1 (db2,db3) = 1                        */

    db1 = (a&d&~(b|c|e|f));
    db2 = (b&e&~(a|c|d|f));
    db3 = (c&f&~(a|b|d|e));

/* rest particle and 1 particle */

    ra = (r&a&~(b|c|d|e|f));
    rb = (r&b&~(a|c|d|e|f));
    rc = (r&c&~(a|b|d|e|f));
    rd = (r&d&~(a|b|c|e|f));
    re = (r&e&~(a|b|c|d|f));
    rf = (r&f&~(a|b|c|d|e));

/* no rest particle and 2 particles (i,i+2) */
```

```
    ra2 = (f&b&~(r|a|c|d|e));
    rb2 = (a&c&~(r|b|d|e|f));
    rc2 = (b&d&~(r|a|c|e|f));
    rd2 = (c&e&~(r|a|b|d|f));
    re2 = (d&f&~(r|a|b|c|e));
    rf2 = (e&a&~(r|b|c|d|f));

/* change a and d
   <-> three-body collision triple=1
       or two-body collision db1=1
       or two-body collision db2=1 and eps=1    (- rotation)
       or two-body collision db3=1 and noeps=1 (+ rotation)
   <-> chad=1                                           */

    eps = irn[ix][iy];        /* random bits */
    noeps = ~eps;

cha=(triple|db1|(eps&db2)|(noeps&db3)|ra|rb|rf|ra2|rb2|rf2);
chd=(triple|db1|(eps&db2)|(noeps&db3)|rd|rc|re|rd2|rc2|re2);
chb=(triple|db2|(eps&db3)|(noeps&db1)|rb|ra|rc|rb2|ra2|rc2);
che=(triple|db2|(eps&db3)|(noeps&db1)|re|rd|rf|re2|rd2|rf2);
chc=(triple|db3|(eps&db1)|(noeps&db2)|rc|rb|rd|rc2|rb2|rd2);
chf=(triple|db3|(eps&db1)|(noeps&db2)|rf|ra|re|rf2|ra2|re2);
chr=(ra|rb|rc|rd|re|rf|ra2|rb2|rc2|rd2|re2|rf2);

/* change: a = a ^ chad */

    k1[ix][iy] = i1[ix][iy]^cha;
    k2[ix][iy] = i2[ix][iy]^chb;
    k3[ix][iy] = i3[ix][iy]^chc;
    k4[ix][iy] = i4[ix][iy]^chd;
    k5[ix][iy] = i5[ix][iy]^che;
    k6[ix][iy] = i6[ix][iy]^chf;
    rest[ix][iy] ^= chr;

/* collision finished (except at the boundaries) */
}}
```

```
/*
i=========================================================i
i                                                         i
i      c o l l i s i o n       FHP-III                    i
i      -------------------                                i
i                                                         i
i ( two, three and four body collisions)                 i
i    162 bit-operations                                   i
i=========================================================i


collision with 1 rest particle (FHP-III)
24.6.91 Armin Vogeler                                    */

    for(ix=0; ix < IXM; ix++)
        for(iy=0; iy < IYM; iy++) {


    a = i1[ix][iy];
    b = i2[ix][iy];
    c = i3[ix][iy];
    d = i4[ix][iy];
    e = i5[ix][iy];
    f = i6[ix][iy];
    r = rest[ix][iy];
    s = sb[ix][iy];
    eps = irn[ix][iy];


    ns = ~s;           /* no solid bit */
    h1 = a&c&e;        /* 3 particles a,c,e */
    h2 = b&d&f;        /* 3 particles b,d,f */
    h3 = a^c^e;        /* 1 or 3 particles in a,c,e */
    h4 = b^d^f;        /* 1 or 3 particles in b,d,f */
    h5 = a|c|e;        /* at least 1 particle in a,c,e */
    h6 = b|d|f;        /* at least 1 particle in b,d,f */
    /* 0 particles in a,c,e or!! in b,d,f and no solid */
    h0 = ns&(h5^h6);
    /* 2 particles in a,c,e or b,d,f and no solid */
    z1 = ns&((~h3&h5)^(~h4&h6));

/* three-body collisions */
```

```
    c3 = (h1^h2)&h0;

/* head-on collisions with spectator */

    c2s = z1&((a&d)^(b&e)^(c&f));

/* two- and four-body collisions */

    c24 = ns&((f^a)|(a^b))&(~((f^c)|(a^d)|(b^e)));

/* rest particle and 1 particle collisions */

    r1 = r&((h3&~h1)^(h4&~h2))&h0;

/* no rest particle and 2 particles collisions */

    r2 = z1&h0&~r;

/* no s,c24, r1, r2 and c2s collision */
    no = ~(c24|r1|r2|c2s)&ns;
    le = c24&eps;      /* c24 collision and left  rotation */
    ri = c24&~eps;     /* c24 collision and right rotation */

/* change bitfield: */

/*|---| |----------| |----------| |------| |---------| |-----|*/
/*|   | |          | |   |----|| |      | ||---|   | |     |*/
k1[ix][iy]
=(s&d)|(ri&b^le&f)|(no&(a^c3))|(b&f&r2)|((b^f)&r1)|(~d&c2s);
k2[ix][iy]
=(s&e)|(ri&c^le&a)|(no&(b^c3))|(a&c&r2)|((a^c)&r1)|(~e&c2s);
k3[ix][iy]
=(s&f)|(ri&d^le&b)|(no&(c^c3))|(b&d&r2)|((b^d)&r1)|(~f&c2s);
k4[ix][iy]
=(s&a)|(ri&e^le&c)|(no&(d^c3))|(c&e&r2)|((c^e)&r1)|(~a&c2s);
k5[ix][iy]
=(s&b)|(ri&f^le&d)|(no&(e^c3))|(d&f&r2)|((d^f)&r1)|(~b&c2s);
k6[ix][iy]
=(s&c)|(ri&a^le&e)|(no&(f^c3))|(e&a&r2)|((e^a)&r1)|(~c&c2s);
krest[ix][iy] = r^(r1|r2); }

/*------------             end of collision           ---------- */
```

6.4 Thermal LBM: derivation of the coefficients

Constraints from the definitions of mass and momentum.

Mass:

$$\rho = \sum_i F_i^{eq} = A_0 + 6\left(A_1 + A_2\right) + \left(3C_1 + 12C_2 + D_0 + 6D_1 + 6D_2\right)u^2$$

→ constraints 1 and 2

$$3C_1 + 12C_2 + D_0 + 6D_1 + 6D_2 = 0 \qquad (6.4.1)$$

$$A_0 + 6\left(A_1 + A_2\right) = \rho \qquad (6.4.2)$$

Momentum:

$$j = \sum_i c_i F_i^{eq} = u\left[3B_1 + 12B_2 + \left(\frac{9}{4}E_1 + 36E_2 + 3G_1 + 12G_2\right)u^2\right]$$

→ constraints 3 and 4

$$3B_1 + 12B_2 = \rho \qquad (6.4.3)$$

$$\frac{9}{4}E_1 + 36E_2 + 3G_1 + 12G_2 = 0 \qquad (6.4.4)$$

Conservation of mass, momentum, and energy.

The expansions (5.2.25) will be substituted into the conservation equations for mass, momentum and energy

$$0 = \sum_i \begin{pmatrix} 1 \\ c_i \\ c_i^2/2 \end{pmatrix} \left[F_i\left(x + c_i, t + 1\right) - F_i(x, t)\right]$$

which lead to

$$0 \underset{(5.2.25)}{=} \sum_i \begin{pmatrix} 1 \\ c_{i\alpha} \\ c_{i\alpha}c_{i\alpha}/2 \end{pmatrix} \left[F_i(x,t) + \partial_t F_i + c_{i\beta}\partial_{x_\beta} F_i + \frac{1}{2}\partial_t\partial_t F_i\right.$$

$$\left. + \frac{1}{2}\partial_{x_\beta}\partial_{x_\gamma} c_{i\beta}c_{i\gamma} F_i + c_{i\beta}\partial_t\partial_{x_\beta} F_i + \mathcal{O}(\partial^3 F_i) - F_i(x,t)\right]$$

$$= \sum_i \begin{pmatrix} 1 \\ c_{i\alpha} \\ c_{i\alpha}c_{i\alpha}/2 \end{pmatrix} \left[\epsilon\partial_t^{(1)} F_i^{(0)} + \epsilon^2\partial_t^{(1)} F_i^{(1)} + \epsilon^2\partial_t^{(2)} F_i^{(0)}\right.$$

$$+ c_{i\beta}\epsilon\partial_{x_\beta}^{(1)} F_i^{(0)} + c_{i\beta}\epsilon^2\partial_{x_\beta}^{(1)} F_i^{(1)} + \frac{1}{2}\epsilon^2\partial_t^{(1)}\partial_t^{(1)} F_i^{(0)}$$

$$\left. + \frac{1}{2}\epsilon^2\partial_{x_\beta}^{(1)}\partial_{x_\gamma}^{(1)} c_{i\beta}c_{i\gamma} F_i^{(0)} + c_{i\beta}\epsilon^2\partial_t^{(1)}\partial_{x_\beta}^{(1)} F_i^{(0)} + \mathcal{O}(\epsilon^3)\right]$$

and finally sorted according to orders in ϵ

$$0 = \sum_i \begin{pmatrix} 1 \\ c_{i\alpha} \\ c_{i\alpha}c_{i\alpha}/2 \end{pmatrix} \left\{ \epsilon \left[\partial_t^{(1)} F_i^{(0)} + c_{i\beta} \partial_{x_\beta}^{(1)} F_i^{(0)} \right] \right.$$

$$+ \epsilon^2 \left[\partial_t^{(1)} F_i^{(1)} + \partial_t^{(2)} F_i^{(0)} + c_{i\beta} \partial_{x_\beta}^{(1)} F_i^{(1)} + \frac{1}{2} \partial_t^{(1)} \partial_t^{(1)} F_i^{(0)} \right. \quad (6.4.5)$$

$$+ \frac{1}{2} \partial_{x_\beta}^{(1)} \partial_{x_\gamma}^{(1)} c_{i\beta} c_{i\gamma} F_i^{(0)} + \left. c_{i\beta} \partial_{x_\beta}^{(1)} \partial_t^{(1)} F_i^{(0)} \right] + \mathcal{O}(\epsilon^3) \right\}$$

Terms of first order in ϵ: mass.

To first order in ϵ eq. (6.4.5) yields:

$$0 = \sum_i \left[\partial_t^{(1)} F_i^{(0)} + c_{i\alpha} \partial_{x_\alpha}^{(1)} F_i^{(0)} \right]$$

or

$$\partial_t^{(1)} \rho + \partial_{x_\alpha}^{(1)} j_\alpha = 0 \quad \text{(continuity equation)}, \quad (6.4.6)$$

$\rightarrow$ no further constraints from mass conservation.

Terms of first order in ϵ: momentum.

$$0 = \sum_i \left[c_{i\alpha} \partial_t^{(1)} F_i^{(0)} + \partial_{x_\beta}^{(1)} c_{i\alpha} c_{i\beta} F_i^{(0)} \right]$$

$$0 = \partial_t^{(1)} (\rho u_\alpha) + \partial_{x_\beta}^{(1)} P_{\alpha\beta}^{(0)}. \quad (6.4.7)$$

whereby

$$P_{\alpha\beta}^{(0)} := \sum_i c_{i\alpha} c_{i\beta} F_i^{(0)}$$

is the momentum flux tensor with components

$$P_{xx}^{(0)} = 3A_1 + 12A_2 + \left(\frac{3}{4}C_1 + 12C_2 \right)(3u^2 + v^2) + (3D_1 + 12D_2)u^2$$

$$= 3A_1 + 12A_2 + \left(\frac{9}{4}C_1 + 36C_2 + 3D_1 + 12D_2 \right) u^2$$

$$+ \left(\frac{3}{4}C_1 + 12C_2 + 3D_1 + 12D_2 \right) v^2$$

$$P_{xy}^{(0)} = \left(\frac{3}{2}C_1 + 24C_2\right) uv = P_{yx}^{(0)} \tag{6.4.8}$$

$$P_{yy}^{(0)} = 3A_1 + 12A_2 + \left(\frac{3}{4}C_1 + 12C_2\right)(u^2 + 3v^2) + (3D_1 + 12D_2)\,u^2$$

$$= 3A_1 + 12A_2 + \left(\frac{3}{4}C_1 + 12C_2 + 3D_1 + 12D_2\right) u^2$$

$$+ \left(\frac{9}{4}C_1 + 36C_2 + 3D_1 + 12D_2\right) v^2$$

The momentum flux tensor should yield

$$P_{\alpha\beta}^{(0)} = \rho u_\alpha u_\beta + p\,\delta_{\alpha\beta} = \rho \begin{pmatrix} u^2 & uv \\ uv & v^2 \end{pmatrix} + p\,\delta_{\alpha\beta}. \tag{6.4.9}$$

Comparison of (6.4.8) and (6.4.9) leads to:

$$3A_1 + 12A_2 = p$$

$$\frac{9}{4}C_1 + 36C_2 + 3D_1 + 12D_2 = \rho$$

$$\frac{3}{4}C_1 + 12C_2 + 3D_1 + 12D_2 = 0$$

$$\frac{3}{2}C_1 + 24C_2 = \rho.$$

This results in the three independent constraints 5 to 7:

$$3A_1 + 12A_2 = p \tag{6.4.10}$$

$$\frac{3}{2}C_1 + 24C_2 = \rho \tag{6.4.11}$$

and

$$6D_1 + 24D_2 = -\rho \tag{6.4.12}$$

The first order terms of the moment equation lead to the Euler equation

$$\underbrace{=}_{(6.4.6)} \quad \rho\partial_t^{(1)}u_\alpha - u_\alpha\partial_{x_\beta}^{(1)}(\rho u_\beta)$$

$$= \quad \rho\partial_t^{(1)}u_\alpha + u_\alpha\partial_t^{(1)}\rho$$

$$\partial_t^{(1)}(\rho u_\alpha) \quad =$$

$$= \quad -\partial_{x_\beta}^{(1)}P_{\alpha\beta}^{(0)}$$

$$\underbrace{=}_{(6.4.9)} \quad -\partial_{x_\beta}^{(1)}(\rho u_\alpha u_\beta + p\delta_{\alpha\beta})$$

$$= \quad -\rho u_\beta\partial_{x_\beta}^{(1)}u_\alpha - u_\alpha\partial_{x_\beta}^{(1)}(\rho u_\beta) - \partial_{x_\alpha}^{(1)}p \tag{6.4.13}$$

and therefore

$$\rho \partial_t^{(1)} u_\alpha = -\rho u_\beta \partial_{x_\beta}^{(1)} u_\alpha - \partial_{x_\alpha}^{(1)} p \qquad (6.4.14)$$

Terms of first order in ϵ: energy.

$$0 = \frac{1}{2} \sum_i \left[\partial_t^{(1)} c_{i\alpha} c_{i\alpha} F_i^{(0)} + \partial_{x_\beta}^{(1)} c_{i\alpha} c_{i\alpha} c_{i\beta} F_i^{(0)} \right]$$

$$\frac{1}{2} \partial_t^{(1)} \sum_i c_{i\alpha} c_{i\alpha} F_i^{(0)} = \frac{1}{2} \partial_t^{(1)} P_{\alpha\alpha}^{(0)} = \partial_t^{(1)} \left[\underbrace{\frac{1}{2} \rho u_\alpha u_\alpha}_{= \rho\varepsilon_K} + \underbrace{p}_{= \rho\varepsilon_I} \right]$$

where p is to be identified with the internal energy, i.e. $p = \rho\varepsilon_I$, and $\frac{1}{2}\rho u_\alpha u_\alpha$ is the kinetic energy.

$$\frac{1}{2} \partial_{x_\beta}^{(1)} \sum_i c_{i\alpha} c_{i\alpha} c_{i\beta} F_i^{(0)}$$

$$= \partial_{x_\beta}^{(1)} \left[\underbrace{\left(\frac{3}{2}B_1 + 24B_2\right) u_\beta}_{=: f_1(\rho, \varepsilon_I)} + \underbrace{\left(\frac{9}{8}E_1 + 72E_2 + \frac{3}{2}G_1 + 24G_2\right) u_\alpha u_\alpha u_\beta}_{=: f_2(\rho, \varepsilon_I)} \right]$$

$$= \partial_{x_\beta}^{(1)} \left[f_1(\rho, \varepsilon_I) u_\beta + f_2(\rho, \varepsilon_I) u_\alpha u_\alpha u_\beta \right] \qquad (6.4.15)$$

and thus

$$\partial_t^{(1)}(\rho\varepsilon_K + \rho\varepsilon_I) = -\partial_{x_\beta}^{(1)} \left[f_1(\rho, \varepsilon) u_\beta + f_2(\rho, \varepsilon) u_\alpha u_\alpha u_\beta \right] \qquad (6.4.16)$$

$$
\begin{aligned}
\partial_t^{(1)}(\rho\varepsilon_K) \quad &= \quad \partial_t^{(1)}\left(\frac{1}{2}\rho u_\alpha u_\alpha\right) \\[2mm]
&= \quad \frac{1}{2} u_\alpha \partial_t^{(1)}(\rho u_\alpha) + \frac{1}{2}\rho u_\alpha \partial_t^{(1)} u_\alpha \\[2mm]
\underset{(6.4.13),(6.4.14)}{=} \quad & -\frac{1}{2} u_\alpha [\rho u_\beta \partial_{x_\beta}^{(1)} u_\alpha + u_\alpha \partial_{x_\beta}^{(1)}(\rho u_\beta) + \partial_{x_\alpha}^{(1)} p \\[2mm]
& + \rho u_\beta \partial_{x_\beta}^{(1)} u_\alpha + \partial_{x_\alpha}^{(1)} p] \\[2mm]
= \quad & -\rho u_\alpha u_\beta \partial_{x_\beta}^{(1)} u_\alpha - u_\alpha \partial_{x_\alpha}^{(1)} p \\[2mm]
& -\frac{1}{2} u_\alpha u_\alpha \partial_{x_\beta}^{(1)}(\rho u_\beta) \qquad (6.4.17)
\end{aligned}
$$

$$\rho\partial_t^{(1)}\varepsilon_I \quad = \quad \partial_t^{(1)}(\rho\varepsilon_I) - \varepsilon_I\partial_t^{(1)}\rho$$

$$\underbrace{=}_{\text{(6.4.6), (6.4.16)}} \quad -\partial_t^{(1)}(\rho\varepsilon_K) - \partial_{x_\alpha}^{(1)}[f_1 u_\alpha + f_2 u_\alpha u_\beta u_\beta] + \varepsilon_I\partial_{x_\alpha}^{(1)}(\rho u_\alpha)$$

$$\underbrace{=}_{\text{(6.4.17)}} \quad \rho u_\alpha u_\beta\partial_{x_\beta}^{(1)}u_\alpha + u_\alpha\partial_{x_\alpha}^{(1)}p + \frac{1}{2}u_\alpha u_\alpha\partial_{x_\beta}^{(1)}(\rho u_\beta)$$

$$-\partial_{x_\alpha}^{(1)}[f_1 u_\alpha + f_2 u_\alpha u_\beta u_\beta] + \varepsilon_I\partial_{x_\alpha}^{(1)}(\rho u_\alpha) \qquad (6.4.18)$$

Substitution of $p = \rho\varepsilon_I$ and expansion of all terms leads to

$$\rho\partial_t^{(1)}\varepsilon_I \;=\; \underbrace{\rho u_\alpha u_\beta\partial_{x_\beta}^{(1)}u_\alpha}_{(1)} + \underbrace{\rho u_\alpha\partial_{x_\alpha}^{(1)}\varepsilon_I}_{(2)} + \underbrace{u_\alpha\varepsilon_I\partial_{x_\alpha}^{(1)}\rho}_{(3)}$$

$$+ \underbrace{\frac{1}{2}\rho u_\beta u_\beta\partial_{x_\alpha}^{(1)}u_\alpha}_{(4)} + \underbrace{\frac{1}{2}u_\alpha u_\beta u_\beta\partial_{x_\alpha}^{(1)}\rho}_{(5)}$$

$$- \underbrace{f_1\partial_{x_\alpha}^{(1)}u_\alpha}_{(6)} - \underbrace{u_\alpha\partial_{x_\alpha}^{(1)}f_1}_{(7)}$$

$$- \underbrace{u_\alpha u_\beta u_\beta\partial_{x_\alpha}^{(1)}f_2}_{(8)} - \underbrace{f_2 u_\beta u_\beta\partial_{x_\alpha}^{(1)}u_\alpha}_{(9)} - \underbrace{2f_2 u_\alpha u_\beta\partial_{x_\alpha}^{(1)}u_\beta}_{(10)}$$

$$+ \underbrace{\rho\varepsilon_I\partial_{x_\alpha}^{(1)}u_\alpha}_{(11)} + \underbrace{u_\alpha\varepsilon_I\partial_{x_\alpha}^{(1)}\rho}_{(12)} \qquad (6.4.19)$$

which should yield

$$\rho\partial_t^{(1)}\varepsilon_I = -\underbrace{\rho u_\alpha\partial_{x_\alpha}^{(1)}\varepsilon_I}_{(13)} - \underbrace{\rho\varepsilon_I\partial_{x_\alpha}^{(1)}u_\alpha}_{(14)} \qquad (6.4.20)$$

Terms (1) and (10), (4) and (9), and (5) and (8) cancel each other when $f_2 = \rho/2$. The sum (2) + (3) + (7) + (12) gives (13) when $f_1 = 2\rho\varepsilon_I$. Finally, (6) + (11) gives (14). Thus we obtain the constraints 8 and 9:

$$\frac{3}{2}B_1 + 24B_2 = 2\rho\varepsilon_I \;(= f_1) \qquad (6.4.21)$$

$$\frac{9}{8}E_1 + 72E_2 + \frac{3}{2}G_1 + 24G_2 = \frac{\rho}{2}(=f_2) \tag{6.4.22}$$

Calculation of the coefficients.

B_1 and B_2 are constrained by (6.4.3) and (6.4.21)

$$3B_1 + 12B_2 = \rho$$

$$\frac{3}{2}B_1 + 24B_2 = 2\rho\varepsilon_I$$

which lead to the unique solution

$$B_1 = \frac{4}{9}\rho(1 - \varepsilon_I) \tag{6.4.23}$$

$$B_2 = \frac{\rho}{36}(4\varepsilon_I - 1). \tag{6.4.24}$$

A_1 and A_2 are constrained only by (6.4.2) and (6.4.10) and A_0 only by (6.4.2):

$$A_0 + 6A_1 + 6A_2 = \rho$$

$$3A_1 + 12A_2 = p(= \rho\varepsilon_I).$$

In order to obtain a unique solution one may require (compare a similar constraint in Section 5.4, Eq. 5.4.4) that $A_1/B_1 = A_2/B_2$. This leads to

$$A_0 = \rho(1 - \frac{5}{2}\varepsilon_I + 2\varepsilon_I^2), \quad A_1 = \rho\frac{4}{9}(\varepsilon_I - \varepsilon_I^2), \quad A_2 = \rho\frac{1}{36}(-\varepsilon_I + 4\varepsilon_I^2),$$

e.g. the expressions given by Alexander et al. (1993).

The C_ν and D_ν are constrained by (6.4.1), (6.4.11), and (6.4.12):

$$3C_1 + 12C_2 + D_0 + 6D_1 + 6D_2 = 0$$

$$\frac{3}{2}C_1 + 24C_2 = \rho$$

and

$$6D_1 + 24D_2 = -\rho.$$

The choice of Alexander et al. is consistent with these constraints.

E_ν and G_ν are constrained by (6.4.4) and (6.4.22):

$$\frac{9}{4}E_1 + 36E_2 + 3G_1 + 12G_2 = 0$$

$$\frac{9}{8}E_1 + 72E_2 + \frac{3}{2}G_1 + 24G_2 = \frac{\rho}{2}.$$

Alexander et al. (1993) have chosen $G_1 = 0 = G_2$. Then E_1 and E_2 are uniquely given by

$$E_1 = -\frac{4}{27}\rho \tag{6.4.25}$$

$$E_2 = \frac{\rho}{108} \tag{6.4.26}$$

6.5 Schläfli symbols

Regular polytopes can be characterized by Schläfli[2] symbols instead of listing, for example, the coordinates of the whole set of vertices. Coxeter (1963, p.126/7) defines a polytope "as a finite convex[3] region of n-dimensional space enclosed by a finite number of hyperplanes". A polytope is characterized by its ensemble of *vertices*. Two-dimensional polytopes are called *polygons*. Three-dimensional polytopes are called *polyhedra*. The part of the polytope that lies in one of the hyperplanes is called a *cell* (each cell is a $(n-1)$-dimensional polytope; example: consider the cube where the cells are squares). The cells of polyhedra are called *faces*; they are polygons bounded by *edges* or *sides*. Edges join nearest-neighbor vertices. Thus a four-dimensional polytope Π_4 has solid cells Π_3, plane faces Π_2 (separating two cells), edges Π_1, and vertices Π_0.

A polygon with p vertices is said to be *regular* if it is both *equilateral* (all sides are equal) and *equiangular* (all angles between nearest neighbor vertices are equal). If $p > 3$ a polygons can be equilateral without being equiangular (a rhomb, for example), or vice versa (a rectangle). Regular polygons are denoted by $\{p\}$ (the Schläfli symbol = number of vertices put in cranked brackets); thus $\{3\}$ is an equilateral triangle, $\{4\}$ is a square, $\{5\}$ is a regular pentagon, and so on.

A polyhedron is said to be *regular* if its faces are regular and equal, while its vertices are all surrounded alike. If its faces are $\{p\}$'s (i.e. regular polygons), q surrounding each vertex, the polyhedron is denoted by the Schläfli symbol $\{p, q\}$. In three dimensions there exist only five regular polyhedra, namely the Platonic solids (compare Section 3.4). Consider, for instance, the cube. The faces are squares (4 edges) and each vertex is surrounded by 3 faces. Accordingly the cube is denoted by the Schläfli symbol $\{4, 3\}$. The Schläfli symbols for the other Platonic solids read: tetrahedron $\{3, 3\}$, octahedron $\{3, 4\}$, dodecahedron $\{5, 3\}$, icosahedron $\{3, 5\}$. Please note that the dual polytope to $\{p, q\}$ has the Schläfli symbol $\{q, p\}$.

A polytope Π_n ($n > 2$) is said to be *regular* if its cells are regular and there is a regular *vertex figure*[4] at every vertex ('regular surrounded'). It can be shown that as a consequence of this definition all cells are equal ($\{p, q\}$ for $n = 4$) and the vertex figures are all equal ($\{q, r\}$ for $n = 4$). A regular polytope Π_4 is denoted by the Schläfli symbol $\{p, q, r\}$ where the cells are $\{p, q\}$ and r is the number of cells that surround an edge. The three regular polytopes in four dimensions

---

[2] After the Swiss mathematician Ludwig Schläfli (1814-95).

[3] "A region is said to be **convex** if it contains the whole of the segment joining every pair of its points." (Coxeter, 1963, p.126)

[4] "If the mid-points of all the edges that emanate from a given vertex O of Π_n lie in one hyperplane ..., then these mid-points are the vertices of an $(n-1)$-dimensional polytope called the *vertex figure* of Π_n at O." (Coxeter, 1963, p. 128)

$$\{3,3,3\}, \quad \{3,3,4\}, \quad \{4,3,3\}, \tag{6.5.1}$$

are bounded by tetrahedrons ($\{3,3\}$) or cubes ($\{4,3\}$). $\Pi_4 = \{4,3,3\}$ is the face-centered hypercube (FCHC). Similarly, a regular polytope Π_5 whose cells are $\{p,q,r\}$ must have vertex figures $\{q,r,s\}$, and thus will be denoted by

$$\Pi_5 = \{p,q,r,s\}. \tag{6.5.2}$$

It can be shown (Coxeter, 1963) that the parameters of the Schläfli for regular polyhedra are constrained by "Schläfli's criterion" which reads

$$\frac{1}{p} + \frac{1}{q} > \frac{1}{2} \quad \text{for} \quad \{p,q\} \tag{6.5.3}$$

and

$$\sin\frac{\pi}{p}\sin\frac{\pi}{r} > \sin\frac{\pi}{q} \quad \text{for} \quad \{p,q,r\}. \tag{6.5.4}$$

This and $p,q,r \geq 3$ leads to

$$\{3,3\}, \quad \{3,4\}, \quad \{4,3\}, \quad \{3,5\}, \quad \{5,3\} \tag{6.5.5}$$

and

$$\{3,3,3\}, \quad \{3,3,4\}, \quad \{4,3,3\}, \quad \{3,4,3\}, \quad \{3,3,5\}, \quad \{5,3,3\}. \tag{6.5.6}$$

Since Schläfli's criterion is merely a *necessary* condition, it remains to be proved that the corresponding polytopes actually exist (this can be very laboriously!; result: all above mentioned polytopes exist except $\{3,4,3\}$, $\{3,3,5\}$, $\{5,3,3\}$).

6.6 Notation, symbols and abbreviations

General remarks:
Latin indices refer to the lattice vectors and run from 0 or 1 to l where l is the number of non-vanishing lattice velocities.

Greek indices assign the cartesian components of vectors and therefore run from 1 to D where D is the dimension.

If not otherwise stated (Einstein's) summation convention is used, i.e. summation is performed over repeated indices ($n_i c_i = \sum_{i=1}^{l} n_i c_i$). No summation will be done over primed indices ($n_{j'} v_{j'} \neq \sum_{j'=1}^{l} n_{j'} v_{j'}$).

Table 6.6.1. Notation (miscellaneous symbols)

Symbol	Meaning
∇	nabla operator
∂	partial derivative
&	AND (Boolean operator)
|	OR (inclusive or; Boolean operator)
$\wedge$	XOR (exclusive or; Boolean operator)
$\sim$	NOT (Boolean operator)
$\cup$	union (Boolean algebra)
$\cap$	intersection (Boolean algebra)
$^{-}$	complementation (Boolean algebra)
$\circ$	composition (of two elements; group theory)

Table 6.6.2. Notation (Latin letters)

Symbol	Meaning
A	Lagrange multiplier
$A(s \rightarrow s')$	transition probability
$A_{ss'}$	transition matrix
$A_\nu, B_\nu \ldots$	free parameters of equilibrium distributions
$a_i^{(t)}$	state of cell i at time t
B	Boolean algebra
B	Lagrange multiplier
b	number of lattice velocities ('bits')
C	collision operator
C	number of corners
c_i	lattice vectors, lattice velocities
$c_{i\alpha}$	cartesian component of the ith lattice velocity
c_s	speed of sound
c_v	heat capacity at constant volume
D	(spatial) dimension
$DkQb$	lattice notation (k=dimension, b=number of lattice velocities)
d	density per cell
$\mathcal{E}$	evolution operator
E	number of edges
E_A	Ekman number
$\mathcal{F}$	operator that interchanges particles and holes
F	body force
F	number of faces
F_m	distribution functions
f	Coriolis parameter
f_0	Coriolis parameter at φ_0
$\mathcal{G}$	isometric group
$G_{\alpha_1\alpha_2\ldots\alpha_n}$	generalized lattice tensor of rank n
$G(\rho)$	g-factor (breaking Galilean invariance)
$g(\rho)$	$= G(\rho)/2$; g-factor (breaking Galilean invariance)
g	group element
H	Boltzmann's H (= - entropy)
H_i	Zanetti invarinats
h	Lagrange multiplier
$\mathcal{I}$	identity operator
$I(P)$	measure of information
$J(f)$	collision operator (BGK)
$J(a, b)$	Jacobi operator
j	momentum density
K_n	Knudsen number
k	number of states (CA)
k_B	Boltzmann constant
k_s	friction coefficient

Table 6.6.3. Notation (Latin letters; continued)

Symbol	Meaning
$\mathcal{L}$	a lattice
L	characteristic length scale
$L_{\alpha_1\alpha_2\ldots\alpha_n}$	lattice tensor of rank n
M_a	Mach number
$\mathbb{N}$	the set of natural numbers (integers)
$\mathbb{N}_0$	the set of non-negative integers
N_i	mean occupation number (real variable)
n_i	occupation number (Boolean variable)
$\mathcal{O}(\epsilon^2)$	on the order of ϵ^2
$O_{\alpha\beta}$	orthogonal transformation matrix
P	discrete probability distribution
P	kinematic pressure
$P_{\alpha\beta}$	momentum flux tensor
p	pressure
p_i	probabilities
Q	set of possible automata states
$Q(f,f)$	collision integral
$Q_{i\alpha\beta}$	Q-tensors (FHP)
q	(vectorial) Lagrange multiplier
$\mathbb{R}$	the set of real numbers
R_e	Reynolds number
$R_{e,g}$	grid Reynolds number
R_o	Rossby number
r	range (CA)
r_j	cartesian coordinates of nodes
S	streaming (propagation) operator
S	entropy
T	temperature
$T^{(MA)}_{\alpha\beta\gamma\delta}$	momentum advection tensor (MAT)
$T_{x,y}$	components of the wind stress
t	time
U	characteristic speed
$u = (u, v, w)$	velocity
W_M	Munk scale
W_i	global equilibrium distributions
w_i	weights (generalized lattice tensors)
$x = (x, y, z)$	cartesian coordinates
$\mathbb{Z}_k$	residue class (integers modulo k)
Z	discrete set

Table 6.6.4. Notation (Greek letters)

Symbol	Meaning
β	gradient of the Coriolis parameter
Γ	phase space
Δ_1	collision function
Δt	time step
Δx	spatial step size
$\delta_{\alpha\beta}$	Kronecker symbol
$\delta_{\alpha\beta\gamma\delta}$	generalized Kronecker symbol
ϵ	expansion parameter
$\epsilon_{\alpha\beta\gamma}$	Levi-Civita symbol
ε_I	internal energy density
ε_K	kinetic energy density
θ	$= k_B T$ temperature in energy units
κ	diffusion coefficient
κ	magnitude of wave number
λ	mean free path
ν	shear viscosity
ξ	bulk viscosity
ξ	random Boolean variable
ρ	mass density
σ	collision cross section
τ	collision time
ψ	stream function
ψ_n	collision invariants
Ω	set of events
Ω	angular velocity of the Earth
Ω_i	collision operator
ω	SOR or viscosity parameter

Table 6.6.5. Abbreviations

Acronym	Meaning
BC	boundary conditions
BGK	Bhatnagar, Gross, Krook
CA	cellular automata
EFD	explicit finite difference
FCHC	face-centered hypercube
FHP	Frisch, Hasslacher, Pomeau
HPP	Hardy, Pazzis, Pomeau
LBM	lattice Boltzmann model
LBGK	lattice BGK models
LGCA	lattice-gas cellular automata
MD	molecular dynamics
ODE	ordinary differential equation
PCLBM	pressure corrected lattice Boltzmann model
PDE	partial differential equation
PI	pair interaction
MSC	multi-spin coding
NSE	Navier-Stokes equation
q.e.d.	quot erat demonstrandum
SOR	successive over-relaxation

Index

References

1. Adler, C., B.M. Boghosian, E. G. Flekkøy, N. Margolus and D.H. Rothman. Simulating three-dimensional hydrodynamics on a cellular-automata machine. *J. Stat. Phys.*, 81:105–128, 1995.

2. Adrover, A. and M. Giona. Hydrodynamic properties of fractals: application of the lattice Boltzmann equation to transverse flow past an array of fractal objects. *International Journal of Multiphase Flow*, 23(1):25–35, 1997.

3. Aggarwal, S. Local and global Garden of Eden theorems. Michigan University technical rept. 147, 1973.

4. Aharonov, E. and D.H. Rothman. Non-newtonian flow (through porous media): a lattice-Boltzmann method. *Geophys. Res. Lett.*, 20(8):679–682, 1993.

5. Ahlrichs, P. and B. Dunweg. Lattice Boltzmann simulation of polymer-solvent systems. *International Journal of Modern Physics C*, 9(8):1429–1438, 1998.

6. Aidun, C.K., Yannan Lu and E.-J. Ding. Direct analysis of particulate suspensions with inertia using the discrete Boltzmann equation. *Journal of Fluid Mechanics*, 373:287–311, 1998.

7. Alasyev, V., A. Krasnoproshina, and V. Kryschanovskii. Unsolved problems in homogeneous structures. *Lecture Notes in Computer Science*, 342:33–49, 1989.

8. Alexander, F.J., H. Chen, S. Chen and G.D. Doolen. Lattice Boltzmann model for compressible fluids. *Phys. Rev. A*, 46:1967–1970, 1992.

9. Alexander, F.J., S. Chen and J.D. Sterling. Lattice Boltzmann thermohydrodynamics. *Phys. Rev. E*, 47:R2249–R2252, 1993.

10. Alstrøm, P. and J. Leão. Self-organized criticality in the "game of Life". *Phys. Rev. E*, 49(4):R2507–8, 1994.

11. Amati, G., S. Succi, and R. Benzi. Turbulent channel flow simulations using a coarse-grained extension of the lattice Boltzmann method. *(preprint)*, 1996.

12. Amati, G., S. Succi, and R. Benzi. Turbulent channel flow simulations using a coarse-grained extension of the lattice Boltzmann method. *Fluid Dynamics Research*, 19(5):289–302, 1997a.

13. Amati, G., S. Succi and R. Piva. Massively parallel lattice-Boltzmann simulation of turbulent channel flow. *International Journal of Modern Physics C*, 8(4):869–877, 1997b.

14. Ames, W. F. *Numerical Methods for Partial Differential Equations.* Academic Press, New York, 1977.

15. Amoroso, S. and Y.N. Patt. Decision procedure for surjectivity and injectivity of parallel maps for tesselation structures. *Journal of Computer and System Sciences,* 6:448–464, 1972.

16. Angelopoulos, A.D., V.N. Paunov, V.N. Burganos and A.C. Payatakes. Lattice Boltzmann simulation of nonideal vapor-liquid flow in porous media. *Physical Review E,* 57(3):3237–3245, 1998.

17. Bahr, D.B. and J.B. Rundle. Theory of lattice Boltzmann simulations of glacier flow. *J. Glaciology,* 41(139):634–640, 1995.

18. Bak, Per. *How Nature Works: The Science of Self-Organized Criticality.* Copernicus/Springer, 205pp, 1996.

19. Balasubramanian, K., F. Hayot and W.F. Saam. Darcy's law from lattice-gas hydrodynamics. *Phys. Rev. A,* 36:2248–2253, 1987.

20. Balazs, N.L., B.R. Schlei, and D. Strottman. Relativistic flows on a spacetime lattice. *Acta Physica Hungarica, New Series, Heavy Ion Physics,* 9(1):67–97, 1999.

21. Bandini, S., R. Casati, M. Castagnoli, M. Costato, M. Liguori, and M. Milani. Cellular automata approach for investigation of low power light effects on the dynamics of plant-inhabiting mites. *Nuovo Cimento D,* 20D(10):1595–1607, 1998.

22. Barlovic, R., L. Santen, A. Schadschneider, and M. Schreckenberg. Metastable states in cellular automata for traffic flow. *European Physical Journal B,* 5(3):793–800, 1998.

23. Bartoloni, A., C. Battista, S. Cabasino, P.S. Paolucci, J. Pech, R. Sarno, G.M. Todesco, M. Torelli, W. Tross, P. Vicini, R. Benzi, N. Cabibbo, F. Massaioli, and R. Tripiccione. LBE simulations of Rayleigh-Bénard convection on the APE100 parallel processor. *Int. J. Mod. Phys. C,* 4:993ff, 1993.

24. Bastolla, U. and G. Parisi. Relevant elements, magnetization and dynamical properties in Kauffman networks. A numerical study. *Physica D,* 115(3-4):203–218, 1998a.

25. Bastolla, U. and G. Parisi. The modular structure of Kauffman networks. *Physica D,* 115(3-4):219–233, 1998b.

26. Batchelor, G. K. *An Introduction to Fluid Dynamics.* Cambridge University Press, 1967.

27. Bender, Carl M. and Steven A. Orszag. *Advanced Mathematical Methods for Scientists and Engineers.* McGraw-Hill, Auckland, (2nd printing 1984), 1978.

28. Benjamin, S.C., N.F. Johnson, and P.M. Hui. Cellular automata models of traffic flow along a highway containing a junction. *Journal of Physics A,* 29(12):3119–3127, 1996.

29. Benzi, R., M.V. Struglia, and R. Tripiccione. Extended self-similarity in numerical simulations of three-dimensional anisotropic turbulence. *Physical Review E*, 53(6A):R5565-8, 1996.

30. Benzi, R., R. Tripiccione, F. Massaioli, S. Succi and S. Ciliberto. On the scaling of the velocity and temperature structure functions in Rayleigh-Bénard convection. *Europhys. Lett.*, 25(5):341-346, 1994.

31. Benzi, R., S. Succi and M. Vergassola. The lattice Boltzmann equation: theory and applications. *Phys. Rep.*, 222(3):145-197, 1992.

32. Berest, P., N. Rakotomalala, J.P. Hulin, and D. Salin. Experimental and numerical tools for miscible fluid displacements studies in porous media with large heterogeneities. *European Physical Journal, Applied Physics*, 6(3):309-321, 1999.

33. Berlekamp, E. R., J. H. Conway and R. K. Guy. *Winning Ways for Your Mathematical Plays, 2.* Academic Press, New York, 1984.

34. Bernardin, D. Global invariants and equilibrium states in lattice gases. *J. Stat. Phys.*, 68(3/4):457-495, 1992.

35. Bernsdorf, J., F. Durst, and M. Schäfer. Comparison of cellular automata and finite volume techniques for simulation of incompressible flows in complex geometries. *International Journal for Numerical Methods in Fluids*, 29(3):251-264, 1999.

36. Bernsdorf, J., Th. Zeiser, G. Brenner, and F. Durst. Simulation of a 2D channel flow around a square obstacle with lattice-Boltzmann (BGK) automata. *International Journal of Modern Physics C*, 9(8):1129-1141, 1998.

37. Bhatnagar, P., E. P. Gross and M. K. Krook. A model for collision processes in gases. I. Small amplitude processes in charged and neutral one-component systems. *Phys. Rev.*, 94(3):511-525, 1954.

38. Biggs, M.J. and S.J. Humby. Lattice-gas automata methods for engineering. *Chemical Engineering Research & Design*, 76(A2):162-174, 1998.

39. Binder, P.-M. Topological classification of cellular automata. *Journal of Physics*, 24:L31-L34, 1991.

40. Boghosian, B.M. and C.D. Levermore. A cellular automaton for Burgers' equation. *Complex Systems*, 1:17-30, 1987.

41. Boghosian, B.M. and P.V. Coveney. Inverse Chapman-Enskog derivation of the thermohydrodynamic lattice-BGK model for the ideal gas. *International Journal of Modern Physics C*, 9(8):1231-1245, 1998.

42. Boghosian, B.M. and W. Taylor. Correlations and renormalization in lattice gases. *Physical Review E*, 52(1):510-554, 1995.

43. Boghosian, B.M. and W. Taylor. Quantum lattice-gas models for the many-body Schrödinger equation. *International Journal of Modern Physics C*, 8(4):705-716, 1997.

44. Boghosian, B.M. and W. Taylor, IV. Quantum lattice-gas model for the many-particle Schrodinger equation in d dimensions. *Physical Review E*, 57(1):54-66, 1998.

45. Boghosian, B.M., J. Yepez, F.J. Alexander and N.H. Margolus. Integer lattice gases. *Physical Review E*, 55(4):4137–4147, 1997.

46. Bogoliubov, N.N. Problems of a dynamical theory in statistical mechanics. In J. de Boer and G.E. Uhlenbeck, editor, *Sudies in Statistical Mechanics, Vol. 1*, pages 5–118. North-Holland, Amsterdam, 1962.

47. Boltzmann, L. Weitere Studien über das Wärmegleichgewicht unter Gasmolekülen. *Wien. Ber.*, 66:275–370, 1872.

48. Boon, J. P. and S. Yip. *Molecular Hydrodynamics*. Dover, New York, 1991.

49. Boon, J.P., D. Dab, R. Kapral, and A. Lawniczak. Lattice gas automata for reactive systems. *Physics Reports*, 273(2):55–147, 1996.

50. Bosl, W.J., J. Dvorkin and A. Nur. A study of porosity and permeability using a lattice Boltzmann simulation. *Geophysical Research Letters*, 25(9):1475–1478, 1998.

51. Bourke, W. Spectral methods in global climate and weather prediction models. In Schlesinger, M. E., editor, *Physically Based Modeling and Simulation of Climate and Climate Change, Part 1*, pages 169–220. Kluwer Academic Publishers, 1988.

52. Brankov, J.G.; Priezzhev, V.B.; Schadschneider, A. and Schreckenberg, M. The Kasteleyn model and a cellular automaton approach to traffic flow. *Journal of Physics A*, 29(10):L229–235, 1996.

53. Bryan, K. A numerical method for the study of the circulation of the world ocean. *J. Comput. Phys.*, 4:347–376, 1969.

54. Buckles, J.J., R.D. Hazlett, S. Chen, K.G. Eggert, D.W. Grunau and W.E. Soll. Towards improved prediction of reservoir flow performance. *Los Alamos Science*, 22:112–121, 1994.

55. Buick, J.M. Numerical simulation of internal gravity waves using a lattice gas model. *International Journal for Numerical Methods in Fluids*, 26(6):657–676, 1998.

56. Buick, J.M. and C.A. Greated. Lattice Boltzmann modeling of interfacial gravity waves. *Physics of Fluids*, 10(6):1490–1511, 1998.

57. Buick, J.M., C.A. Greated and D.M. Campbell. Lattice BGK simulation of sound waves. *Europhysical Letters*, 43(3):235–240, 1998.

58. Burks, A.W., editor. *Essays on Cellular Automata*. University of Illinois Press, Urbana, 1970.

59. Burnett, D. The distribution of velocities in a slightly non-uniform gas. *Proc. London Math. Soc.*, 39:385–430, 1935.

60. Burnett, D. The distribution of molecular velocities and the mean motion in a non-uniform gas. *Proc. London Math. Soc.*, 40:382–435, 1936.

61. Bussemaker, H.J. Analysis of a pattern-forming lattice-gas automaton: mean-field theory and beyond. *Physical Review E*, 53(2):1644–1661, 1996.

62. Cancelliere, A., C. Chang, E. Foti, D.H. Rothman, and S. Succi. The permeability of a random medium: Comparison of simulation with theory. *Phys. Fluids A*, 2:2085–2088, 1990.

63. Cao, N., S. Chen, S. Jin, and D. Martinez. Physical symmetry and lattice symmetry in the lattice Boltzmann method. *Physical Review E*, 55(1):R21–R24, 1997.

64. Case, J., D.S. Rajan, and A.M. Shende. Optimally representing euclidean-space discretely for analogically simulating physical phenomena. *Lecture Notes in Computer Science*, 472:190–203, 1990.

65. Cattaneo, G., P. Flocchini, P. and C. Quaranta Vogliotti. An effective classification of elementary cellular automata. In Costato, M.; Degasperis, A.; Milani, M., editor, *Conference Proceedings. National Workshop on Nonlinear Dynamics. Vol.48*, pages 69–78. Bologna, Italy: Italian Phys. Soc, 1995.

66. Cercignani, C. *The Boltzmann Equation and Its Applications*. Springer, New York, 1988.

67. Cercignani, C. *Mathematical Methods in Kinetic Theory*. Plenum, 1990.

68. Chapman, S. On the law of distribution of molecular velocities, and on the theory of viscosity and thermal conduction, in a non-uniform simple monatomic gas. *Phil. Trans. Roy. Soc.*, A216:279–348, 1916.

69. Chapman, S. On the kinetic theory of a gas. Part II. - A composite monatomic gas: Diffusion, viscosity, and thermal conduction. *Phil. Trans. Roy. Soc.*, A217:115–197, 1918.

70. Chapman, S. and T. G. Cowling. *The Mathematical Theory of Non-Uniform Gases*. Cambridge University Press, 1970.

71. Chen, H. H-theorem and generalized semi-detailed balance condition for lattice gas systems. *J. Stat. Phys.*, 81(1/2):347–360, 1995.

72. Chen, H. Entropy, fluctuation and transport in lattice gas systems with generalized semi-detailed balance. *Journal of Plasma Physics*, 57(1):175–186, 1997.

73. Chen, H. Volumetric formulation of the lattice Boltzmann method for fluid dynamics: Basic concept. *Physical Review E*, 58(3):3955–3963, 1998.

74. Chen, H. and W. H. Matthaeus. New cellular automaton model for magneto-hydrodynamics. *Phys. Rev. Lett.*, 58(18):1845–1848, 1987.

75. Chen, H., C. Teixeira and K. Molvig. Digital Physics approach to computational fluid dynamics: some basic theoretical features. *International Journal of Modern Physics C*, 8(4):675–684, 1997.

76. Chen, H., C. Teixeira, and K. Molvig. Realization of fluid boundary conditions via discrete Boltzmann dynamics. *International Journal of Modern Physics C*, 9(8):1281–1292, 1998b.

77. Chen, H., W. H. Matthaeus and L. W. Klein. Theory of multicolor lattice gas: A cellular automaton Poisson solver. *J. Comput. Physics*, 88:433–466, 1990.

78. Chen, S. and G.D. Doolen. Lattice Boltzmann method for fluid flows. *Ann. Rev. Fluid Mech.*, 30:329–364, 1998.

79. Chen, S., D. Martinez, and R. Mei. On boundary conditions in lattice Boltzmann methods. *Physics of Fluids*, 8(9):2527–36, 1996.

80. Chen, S., D.O. Martinez, W.H. Matthaeus and H. Chen. Magnetohydrodynamics computations with lattice gas automata. *J. Stat. Phys.*, 68(3/4):533–556, 1992.

81. Chen, S., G. D. Doolen, and W. H. Matthaeus. Lattice gas automata for simple and complex fluids. *J. Stat. Phys.*, 64(5/6):1133–1162, 1991.

82. Chen, S., H. Chen, D. Martinez, and W.H. Matthaeus. Lattice Boltzmann model for simulation of magnetohydrodynamics. *Phys. Rev. Lett.*, 67(27):3776–3779, 1991.

83. Chen, S., K. Diemer, G.D. Doolen, K. Eggert, C. Fu, S. Gutman, and B.J. Travis. Lattice gas automata for flow through porous media. *Physica D*, 47:72–84, 1991.

84. Chen, S., M. Lee, K.H. Zhao and G.D. Doolen. A lattice gas model with temperature. *Physica D*, 37:42–59, 1989.

85. Chen, S., S.P. Dawson, G.D. Doolen, D.R. Janecky and A. Lawniczak. Lattice methods and their application to reacting systems. *Comp. Chem. Eng.*, 19(6/7):617–646, 1995.

86. Chen, S., Z. Wang, X. Shan and G. Doolen. Lattice Boltzmann computational fluid dynamics in three dimensions. *J. Stat. Phys.*, 68(3/4):379–400, 1992.

87. Chen, Y. and H. Ohashi. Lattice-BGK methods for simulating incompressible fluid flows. *International Journal of Modern Physics C*, 8(4):793–803, 1997.

88. Chen, Y. and H. Ohashi. Lattice Boltzmann method: fundamentals and applications. *Journal of the Japan Society for Simulation Technology*, 17(3):213–219, 1998.

89. Chen, Y., H. Ohashi, and M. Akiyam. A simple method to change the Prandtl number for thermal lattice BGK model. In Tentner, A., editor, *High Performance Computing Symposium 1995 'Grand Challenges in Computer Simulation'*, pages 165–8. Proceedings of the 1995 Simulation Multiconference, San Diego, CA, USA, xxiii+566 pp., 1995.

90. Chen, Y., H. Ohashi and M. Akiyama. Thermal lattice Bhatnagar Gross Krook model without nonlinear deviations in macrodynamic equations. *Phys. Rev. E*, 50(4):2776–2783, 1994.

91. Chen, Y., H. Ohashi, and M. Akiyama. Heat transfer in lattice BGK modeled fluid. *J. Stat. Phys.*, 81(1/2):71–85, 1995.

92. Chen, Y., S. Teng, T. Shukuwa, and H. Ohashi. Lattice-Boltzmann simulation of two-phase fluid flows. *International Journal of Modern Physics C*, 9(8):1383–1391, 1998a.

93. Chenghai, S. Lattice-Boltzmann models for high speed flows. *Physical Review E*, 58(6):7283–7287, 1998.

94. Chopard, B. and M. Droz. Cellular automata model for heat conduction in a fluid. *Phys. Lett. A*, 126(8/9):476, 1988.

95. Chopard, B. and M. Droz. Cellular automata model for the diffusion equation. *J. Stat. Phys.*, 64(3/4):859–892, 1991.

96. Chopard, B. and P.O. Luthi. Lattice Boltzmann computations and applications to physics. *Theoretical Computer Science*, 217(1):115–130, 1999.

97. Chopard, B., P.O. Luthi, and P.-A. Queloz. Cellular automata model of car traffic in a two-dimensional street network. *Journal of Physics A*, 29(10):2325–2336, 1996.

98. Codd, E.F. *Cellular Automata*. Academic Press, New York, 1968.

99. Colvin, M.E., A.J.C. Ladd, and B.J. Alder. Maximally discretized molecular dynamics. *Phys. Rev. Lett.*, 61(4):381–388, 1988.

100. Cornubert, R., D. d'Humières and D. Levermore. A Knudsen layer theory for lattice gases. *Physica D*, 47:241–259, 1991.

101. Courant, R., K. Friedrichs und H. Lewy. Über die partiellen Differentialgleichungen der mathematischen Physik. *Math. Ann.*, 100:32–74, 1928.

102. Coveney, P.V., J.-B. Maillet, J.L. Wilson, P.W. Fowler, O. Al-Mushadani, and B.M. Boghosian. Lattice-gas simulations of ternary amphiphilic fluid flow in porous media. *International Journal of Modern Physics C*, 9(8):1479–90, 1998.

103. Cover, T.M. and J.A. Thomas. *Elements of Information Theory*. 576 pp, John Wiley, New York, 1991.

104. Coxeter, H.S.M. *Regular Polytopes*. Macmillan, 1963.

105. Creutz, M., L. Jacobs and C. Rebbi. Experiments with a gauge-invariant Ising system. *Phys. Rev. Lett.*, 42(21):1390–1393, 1979.

106. Dardis, O. and J. McCloskey. Lattice Boltzmann scheme with real numbered solid density for the simulation of flow in porous media. *Physical Review E*, 57(4):4834–4837, 1998a.

107. Dardis, O. and J. McCloskey. Permeability porosity relationships from numerical simulations of fluid flow. *Geophysical Research Letters*, 25(9):1471–1474, 1998b.

108. Dawson, S.P., S. Chen and G.D. Doolen. Lattice Boltzmann computations for reaction-diffusion equations. *J. Chem. Phys.*, 98(2):1514–1523, 1993.

109. De Fabritiis, G., A. Mancini, D. Mansutti, and S. Succi. Mesoscopic models of liquid/solid phase transitions. *International Journal of Modern Physics C*, 9(8):1405–1415, 1998.

110. de la Torre, A.C. and H.O. Martin. A survey of cellular automata like the 'game of life'. *Journal of Physics A*, 240(3-4):560–570, 1997.

111. Decker, L. and D. Jeulin. Random texture simulation by multi-species lattice gas models. *Journal of Electronic Imaging*, 6(1):78–93, 1997.

112. Deutsch, A. and A.T. Lawniczak. Probabilistic lattice models of collective motion and aggregation: from individual to collective dynamics. *Mathematical Biosciences*, 156(1-2):255–269, 1999.

113. d'Humières, D. and P. Lallemand. Lattice Gas Automata for Fluid Mechanics. *Physica*, 140A:326–335, 1986.

114. d'Humières, D. and P. Lallemand. Numerical Simulations of Hydrodynamics with Lattice Gas Automata in Two Dimensions. *Complex Systems*, 1:599–632, 1987; [Errata: 2,725-726,1989].

115. d'Humières, D., P. Lallemand and J. Searby. Numerical experiments in lattice gases: mixtures and Galilean invariance. *Complex Systems*, 1:633–647, 1987.

116. d'Humières, D., P. Lallemand and U. Frisch. Lattice gas models for 3D hydrodynamics. *Europhys. Lett.*, 2(4):291–297, 1986.

117. d'Humières, D., P. Lallemand and Y.H. Qian. One-dimensional lattice gas models. Divergence of the viscosity. *C.R. Acad. Sci. Paris. Série II*, 308:585–590, 1989.

118. d'Humières, D., Y. H. Qian and P. Lallemand. Finding the linear invariants of lattice gases. In Pires, A., D. P. Landau and H. Hermann, editor, *Computational Physics and Cellular Automata*, pages 97–115. World Scientific, Singapore, 1990.

119. d'Humières, D., Y.H. Qian and P. Lallemand. Invariants in lattice gas models. In R. Monaco, editor, *Discrete Kinetic Theory, Lattice Gas Dynamics and Foundations of Hydrodynamics*, pages 102–113. World Scientific, Singapore, 1989.

120. Doolen, G.D. Bibliography. In G.D. Doolen, U. Frisch, B. Hasslacher, S. Orszag and S. Wolfram, editor, *Lattice Gas Methods for Partial Differential Equations*, pages 509–547. Addison-Wesley, Redwood City, California, 1990.

121. Doolen, G.D., U. Frisch, B. Hasslacher, S. Orszag and S. Wolfram, editor. *Lattice Gas Methods for Partial Differential Equations*. Addison-Wesley, Redwood City, California, 1990.

122. d'Ortona, U., D. Salin, M. Cieplak, R.B. Rybka and J.R. Banavar. Two-color nonlinear Boltzmann cellular automata: Surface tension and wetting. *Physical Review E*, 51(4):3718–3728, 1995.

123. Dubrulle, B. Method of computation of the Reynolds number for two models of lattice gas involving violation of semi-detailed balance. *Complex Systems*, 2:577–609, 1988.

124. Dubrulle, B., M. Hénon, U. Frisch, and J.P. Rivet. Low viscosity lattice gases. *J. Stat. Phys.*, 59(5/6):1187–1226, 1990.

125. Durand, B. and J. Mazoyer. Growing patterns in one dimensional cellular automata. *Complex Systems*, 8:419–434, 1994.

126. Ebihara, K., T. Watanabe, and H. Kaburaki. Surface of dense phase in lattice-gas fluid with long-range interaction. *International Journal of Modern Physics C*, 9(8):1417–1427, 1998.

127. Ekman, V.W. On the influence of the earth's rotation on ocean currents. *Ark. Mat. Astron. Fys.*, 2:1–53, 1905.

128. Ekman, V.W. Über Horizontalzirkulation bei winderzeugten Meeresströmungen. *Ark. Mat. Astron. Fys.*, 17(26):74pp, 1923.

129. Elton, B.H. Comparisons of lattice Boltzmann and finite difference methods for a two-dimensional viscous Burgers equation. *SIAM J. Sci. Comput.*, 17(4):783–813, 1996.

130. Elton, B.H., C.D. Levermore, and G.H. Rodrigue. Lattice Boltzmann methods for some 2-D nonlinear diffusive equations: Computational results. In H. Kaper, editor, *Proceedings of the Workshop on Asymptotic Analysis and Numerical Solution of PDEs (Argonne National Laboratory)*, pages 197–214. Marcel Dekker, New York, 1990.

131. Elton, B.H., C.D. Levermore and G.H. Rodrigue. Convergence of convective-diffusive lattice Boltzmann methods. *SIAM Journal on Numerical Analysis*, 32(5):1327–1354, 1995.

132. Emerton, A.N., P.V. Coveney and B.M. Boghosian. Lattice-gas simulations of domain growth, saturation, and self-assembly in immiscible fluids and microemulsions. *Physical Review E*, 55(1B):708–720, 1997.

133. Emmerich, H., T. Nagatani, and K. Nakanishi. From modified KdV-equation to a second-order cellular automaton for traffic flow. *Physica A*, 254(3-4):548–556, 1998.

134. Enskog, D. *Kinetische Theorie der Vorgänge in mässig verdünnten Gasen.* PhD thesis, Uppsala, 1917.

135. Enskog, D. Kinetische Theorie der Wärmeleitung, Reibung und Selbstdiffusion in gewissen verdichteten Gasen und Flüssigkeiten. *Svensk. Akad. Handl.*, 63(4):5–44, 1922.

136. Ermentrout, G.B. and L. Edlestein-Keshet. Cellular automata approaches to biological modeling. *J. Theor. Biol.*, 160:97–133, 1993.

137. Ernst, M.H. Statistical mechanics of cellular automata fluids. In D. Levesque, J.P. Hansen, and J. Zinn-Justin, editor, *Liquids, Freezing and the Glass Transition, Les Houches 1989*, pages 43–143. Elsevier, Amsterdam, 1991.

138. Evans, D. J. and G. P. Morriss. Nonequilibrium molecular-dynamics simulation of Couette flow in two-dimensional fluids. *Phys. Rev. Lett.*, 51(19):1776–1779, 1983.

139. Fahner, G. A multispeed model for lattice-gas hydrodynamics. *Complex Systems*, 5:1–14, 1991.

140. Fang, H., Z. Lin and Z. Wang. Lattice Boltzmann simulation of viscous fluid systems with elastic boundaries. *Physical Review E*, 57(1):R25–28, 1998.

141. Filippova, O. and D. Hänel. Lattice-Boltzmann simulation of gas-particle flow in filters. *Computers & Fluids*, 26(7):697–712, 1997.

142. Filippova, O. and D. Hänel. Grid refinement for lattice-BGK models. *J. Comput. Phys.*, 147(1):219–228, 1998a.

143. Filippova, O. and D. Hänel. Boundary-fitting and local grid refinement for lattice-BGK models. *International Journal of Modern Physics C*, 9(8):1271–1279, 1998b.

144. Filippova, O. and D. Hänel. Lattice-BGK model for low Mach number combustion. *International Journal of Modern Physics C*, 9(8):1439–1445, 1998c.

145. Flekkøy, E.G. Lattice Bhatnagar-Gross-Krook models for miscible fluids. *Physical Review E*, 47(6):4247–4257, 1993.

146. Flekkøy, E.G. and D.H. Rothman. Fluctuating hydrodynamic interfaces: Theory and simulation. *Physical Review E*, 53(2):1622–1643, 1996a.

147. Flekkøy, E.G., T. Rage, U. Oxaal, and J. Feder. Hydrodynamic irreversibility in creeping flow. *Physical Review Letters*, 77(20):4170–4173, 1996b.

148. Flekkøy, E.G., U. Oxaal, J. Feder and T. Jøssang. Hydrodynamic dispersion at stagnation points: Simulations and experiments. *Phys. Rev. E*, 52(5):4952–4962, 1995.

149. Fogaccia, G., R. Benzi, and F. Romanelli. Lattice Boltzmann algorithm for three-dimensional simulations of plasma turbulence. *Physical Review E*, 54(4):4384–4393, 1996.

150. Fox, A.D. and S.J. Maskell. Two-way interactive nesting of primitive equation ocean models with topography. *J. Phys. Oceanogr.*, 25(12):2977–2996, 1995.

151. Fox, A.D. and S.J. Maskell. A nested primitive equation model of the Iceland-Faeroe front. *J. Geophys. Res.*, 101(C8):18259–18278, 1996.

152. Fredkin, E. Digital mechanics - An informational process based on reversible universal cellular automata. *Physica D*, 45(1-3):254–270, 1990.

153. Freed, D.M. Lattice-Boltzmann method for macroscopic porous media modeling. *International Journal of Modern Physics C*, 9(8):1491–1503, 1998.

154. Friedberg, R. and J.E. Cameron. Test of the Monte Carlo method: fast simulation of a small Ising lattice. *J. Chem. Phys.*, 52:6049–6058, 1970.

155. Frisch, U. Relation between the lattice Boltzmann equation and the Navier-Stokes equations. *Physica D*, 47:231–232, 1991.

156. Frisch, U., B. Hasslacher, and Y. Pomeau. Lattice-gas automata for the Navier-Stokes equation. *Phys. Rev. Lett.*, 56(14):1505–1508, 1986.

157. Frisch, U., D. d'Humières, B. Hasslacher, P. Lallemand, Y. Pomeau and J.-P. Rivet. Lattice gas hydrodynamics in two and three dimensions. *Complex Systems*, 1:649–707, 1987.

158. Fukui, M. and Y. Ishibashi. Traffic flow in 1D cellular automaton model including cars moving with high speed. *Journal of the Physical Society of Japan*, 65(6):1868–1870, 1996a.

159. Fukui, M. and Y. Ishibashi. Effect of reduced randomness on jam in a two-dimensional traffic model. *Journal of the Physical Society of Japan*, 65(6):1871–1873, 1996b.

160. Gallivan, M.A., D.R. Noble, J.G. Georgiadis and R.O. Buckius. An evaluation of the bounce-back boundary condition for lattice Boltzmann simulations. *International Journal for Numerical Methods in Fluids*, 25(3):249–263, 1997.

161. Gardner, M. The fantastic combinations of John Horton Conway's new solitaire game of 'life'. *Sci. Am.*, 223(4):120–123, 1970.

162. Gardner, M. Geometric fallacies: hidden errors pave the road to absurd conclusions. *Sci. Am.*, 224(4):114–117, 1971a.

163. Gardner, M. On cellular automata, self-reproduction, the Garden of Eden and the game "life". *Sci. Am.*, 224(2):112–117, 1971b.

164. Gardner, M. The orders of infinity, the topological nature of dimensions and "supertasks". *Sci. Am.*, 224(3):106–109, 1971c.

165. Gerling, R.W. Classification of triangular and honeycomb cellular automata. *Physica A*, 162:196–209, 1990a.

166. Gerling, R.W. Zellulare Automaten auf dem PC. *Der mathematische und naturwissenschaftliche Unterricht*, 43(8):451–456, 1990b.

167. Gibbs, J. W. *Elementary Principles in Statistical Mechanics*. Yale University Press, New Haven, 1902 (reprinted 1981 by Ox Bow Press, Woodbridge, Connecticut).

168. Ginzbourg, I. and P.M. Adler. Boundary flow conditional analysis for the three-dimensional lattice Boltzmann model. *J. Phys. II France*, 4:191–214, 1994.

169. Ginzbourg, L. and D. d'Humières. Local second-order boundary methods for lattice Boltzmann models. *J. Stat. Phys.*, 1996.

170. Giraud, L. and D. d'Humières. A lattice-Boltzmann model for viscoelasticity. *International Journal of Modern Physics C*, 8(4):805–815, 1997.

171. Giraud, L., D. d'Humières and P. Lallemand. A lattice Boltzmann model for Jeffreys viscoelastic fluid. *Europhysics Letters*, 42(6):625–630, 1998.

172. Goldenfeld, N. and L.P. Kadanoff. Simple lessons from complexity. *Science*, 284:87–89, 1999.

173. Gonnella, G., E. Orlandini and J.M. Yeomans. Lattice-Boltzmann simulations of complex fluids. *International Journal of Modern Physics C*, 8(4):783–792, 1997.

174. Gonnella, G., E. Orlandini and J.M. Yeomans. Lattice Boltzmann simulations of lamellar and droplet phases. *Physical Review E*, 58(1):480–485, 1998.

175. Gonnella, G., E. Orlandini and J.M. Yeomans. Phase separation in two-dimensional fluids: The role of noise. *Physical Review E*, 59(5):R4741–R4744, 1999.

176. Großmann, S. *Funktionalanalysis*. 4., korrigierte Auflage, AULA-Verlag, Wiesbaden, 1988.

177. Grubert, D. Using the FHP-BGK-model to get effective dispersion constants for spatially periodic model geometries. *International Journal of Modern Physics C*, 8(4):817–825, 1997.

178. Gunstensen, A.K. and D.H. Rothman. A Galilean-invariant immiscible lattice gas. *Physica D*, 47:53–63, 1991.

179. Gunstensen, A.K. and D.H. Rothman. Lattice-Boltzmann studies of two-phase flow through porous media. *J. Geophys. Res.*, 98(B4):6431–6441, 1993.

180. Gunstensen, A.K., D.H. Rothman, S. Zaleski and G. Zanetti. Lattice Boltzmann model of immiscible fluids. *Phys. Rev. A*, 43(8):4320–4327, 1991.

181. Gutfraind, R., I. Ippolito and A. Hansen. Study of tracer dispersion in self-affine fractures using lattice-gas automata. *Phys. Fluids*, 7(8):1938–1948, 1995.

182. Gutowitz, H., editor. *Cellular Automata - Theory and Experiment*. MIT Press, Cambridge, Massachusetts, 1991.

183. Hackbusch, W. *Multi-Grid Methods and Applications*. Springer, Berlin, 1985.

184. Halliday, I. Steady state hydrodynamics of a lattice Boltzmann immiscible lattice gas. *Physical Review E*, 53(2):1602–1612, 1996.

185. Halliday, I., C.M. Care, S. Thompson, and D. White. Induced burst of fluid drops in a two-component lattice Bhatnager-Gross-Krook fluid. *Physical Review E*, 54(3):2573–2576, 1996.

186. Halliday, I., S.P. Thompson and C.M. Care. Macroscopic surface tension in a lattice Bhatnagar-Gross-Krook model of two immiscible fluids. *Physical Review E*, 57(1):514–523, 1998.

187. Hardy, J., O. de Pazzis and Y. Pomeau. Molcular dynamics of a lattice gas: Transport properties and time correlation functions. *Phys. Rev.*, A13:1949–1961, 1976.

188. Hardy, J., Y. Pomeau and O. de Pazzis. Time evolution of a two-dimensional model system. I. Invariant states and time correlation functions. *J. Math. Phys.*, 14(12):1746–1759, 1973.

189. Hashimoto, Y. and H. Ohashi. Droplet dynamics using the lattice-gas method. *International Journal of Modern Physics C*, 8(4):977–983, 1997.

190. Hashimoto, Y., Y. Chen, and H. Ohashi. Boundary conditions in lattice gas with continuous velocity. *International Journal of Modern Physics C*, 9(8):1263–1269, 1998.

191. Hasslacher, B. Discrete fluids. Los Alamos Science, 15, special issue, p.175-217, 1987.

192. Hasslacher, B. and D.A. Meyer. Modeling dynamical geometry with lattice-gas automata. *International Journal of Modern Physics C*, 9(8):1597–1605, 1998.

193. Hatori, T. and D. Montgomery. Transport coefficients for magnetohydrodynamic cellular automata. *Complex Systems*, 1:735–752, 1987.

194. Hattori, T. and S. Takesue. Additive conserved quantities in discrete-time lattice dynamical systems. *Physica D*, 49:295–322, 1991.

195. Hayot, F. and L. Wagner. A non-local modification of a lattice Boltzmann model. *Europhysics Letters*, 33(6):435–440, 1996.

196. Hayot, F. and M. R. Lakshmi. Cylinder wake in lattice gas hydrodynamics. *Physica D*, 40:415–420, 1989.

197. He, G. and K. Zhao. Lattice Boltzmann simulation of van der Waals phase transition with chemical potential. *Communications in Theoretical Physics*, 29(4):623–626, 1998.

198. He, X. Error analysis for the interpolation-supplemented lattice-Boltzmann equation scheme. *International Journal of Modern Physics C*, 8(4):737–745, 1997.

199. He, X. and G. Doolen. Lattice Boltzmann method on curvilinear coordinates system: Flow around a circular cylinder. *J. Comput. Phys.*, 134:306–315, 1997a.

200. He, X. and G. Doolen. Lattice Boltzmann method on a curvilinear coordinate system: Vortex shedding behind a circular cylinder. *Physical Review E*, 56(1 A):434–440, 1997b.

201. He, X. and Q. Zou. Analysis and boundary condition of the lattice Boltzmann BGK model with two velocity components. *preprint*, 1995.

202. He, X., L.-S. Luo, and M. Dembo. Some progress in lattice Boltzmann method. Part I. Nonuniform mesh grids. *J. Comput. Physics*, 129(2):357–363, 1996.

203. He, X., Q. Zou, L.-S. Luo and M. Dembo. Analytic solutions of simple flows and analysis of nonslip boundary conditions for the lattice Boltzmann BGK model. *J. Stat. Phys.*, 87(1/2):115–136, 1997.

204. He, X., R. Zhang, S. Chen, and G.D. Doolen. On the three-dimensional Rayleigh-Taylor instability. *Physics of Fluids*, 11(5):1143–1152, 1999b.

205. He, X., S. Chen and G.D. Doolen. A novel thermal model for the lattice Boltzmann method in incompressible limit. *Journal of Computational Physics*, 146(1):282–300, 1998.

206. He, X., S. Chen, and R. Zhang. A lattice Boltzmann scheme for incompressible multiphase flow and its application in simulation of Rayleigh-Taylor instability. *Journal of Computational Physics*, 152(2):642–663, 1999a.

207. Hedrich, R. *Komplexe und fundamentale Strukturen*. Bibliographisches Institut, Mannheim, 1990.

208. Heijs, A.W.J. and C.P. Lowe. Numerical evaluation of the permeability and the Kozeny constant for two types of porous media. *Physical Review E*, 51(5):4346–4352, 1995.

209. Hénon, M. Isometric collision rules for four-dimensional FCHC lattice gas. *Complex Systems*, 1(3):475–494, 1987a.

210. Hénon, M. Viscosity of a Lattice Gas. *Complex Systems*, 1(4):762–790, 1987b.

211. Hénon, M. Optimization of collision rules in the FCHC lattice gas and addition of rest particles. In R. Monaco, editor, *Discrete Kinetic Theory, Lattice Gas Dynamics and Foundations of Hydrodynamics*, pages 146–159. World Scientific, Singapore, 1989a.

212. Hénon, M. On the relation between lattice gases and cellular automata. In R. Monaco, editor, *Discrete Kinetic Theory, Lattice Gas Dynamics and Foundations of Hydrodynamics*, pages 160–161. World Scientific, Singapore, 1989b.

213. Hénon, M. Implementation of the FCHC lattice gas model on the connection machine. *J. Stat. Phys.*, 68(3/4):353–377, 1992.

214. Herrmann, H.J. Simulating granular media on the computer. In Garrido, P.L.; Marro, J., editor, *Third Granada Lectures in Computational Physics. Proceedings of the III. Granada Seminar on Computational Physics*, pages 67–114. Springer-Verlag, Berlin, 1995.

215. Herrmann, H.J.; Flekkøy, E.; Nagel, K.; Peng, G.; Ristow, G. Density waves in dry granular flow. In Wolf, D.E.; Schreckenberg, M.; Bachem, A., editor, *Workshop on Traffic and Granular Flow*, pages 239–250. World Scientific, Singapore, 1996.

216. Heudin, J.-C. A new candidate rule for the game of two-dimensional Life. *Complex Systems*, 10(5):367–381, 1996.

217. Heyes, D.H., G.P. Morriss and D.J. Evans. Nonequilibrium molecular dynamics study of shear flow in soft disks. *J. Chem. Phys.*, 83(9):4760–4766, 1985.

218. Higuera, F. and J. Jiménez. Boltzmann approach to lattice gas simulations. *Europhys. Lett.*, 9(7):663–668, 1989.

219. Higuera, F., S. Succi and R. Benzi. Lattice gas dynamics with enhanced collisions. *Europhys. Lett.*, 9(4):345–349, 1989.

220. Hillis, W.D. Richard Feynman and the Connection Machine. *Physics Today*, 42(2):78–83, 1989.

221. Holdych, D.J., D. Rovas, J.G. Georgiadis, and R.O. Buckius. An improved hydrodynamics formulation for multiphase flow lattice-Boltzmann models. *International Journal of Modern Physics C*, 9(8):1393–1404, 1998.

222. Holme, R. and D. H. Rothman. Lattice-gas and lattice-Boltzmann models of miscible fluids. *J. Stat. Phys.*, 68(3/4):409–429, 1992.

223. Hopcroft, John E. Turing machines. *Sci. Am.*, 250(5):70–80, 1984.

224. Hou, S., Q. Zou, S. Chen, G. Doolen and A.C. Cogley. Simulation of cavity flow by the lattice Boltzmann method. *J. Comput. Physics*, 118(2):329–347, 1995.

225. Hou, S., X. Shan, Q. Zou and G.D. Doolen. Evaluation of two lattice Boltzmann models for multiphase flows. *Journal of Computational Physics*, 138(2):695–713, 1997.

226. Hu, S., G. Yan and W. Shi. A lattice Boltzmann model for compressible perfect gas. *Acta Mechanica Sinica (English Edition)*, 13(3):218–226, 1997.

227. Huang, D.-W. Exact results for car accidents in a traffic model. *Journal of Physics A*, 31(29):6167–6173, 1998.

228. Huang, K. *Statistical Mechanics*. John Wiley & Sons, New York, 1963.

229. Inamuro, T., M. Yoshino and F. Ogino. A non-slip boundary condition for lattice Boltzmann simulations. *Physics of Fluids*, 7(12):2928–2930, 1995; Erratum, 8(4):1124, 1996.

230. Inamuro, T., M. Yoshino, and F. Ogino. Lattice Boltzmann simulation of flows in a three-dimensional porous structure. *International Journal for Numerical Methods in Fluids*, 29(7):737–748, 1999.

231. Ishibashi, Y. and M. Fukui. Phase diagram for the traffic model of two one-dimensional roads with a crossing. *Journal of the Physical Society of Japan*, 65(9):2793–5, 1996.

232. Ishibashi, Y. and M. Fukui. Traverse time in a cellular automaton traffic model. *Journal of the Physical Society of Japan*, 65(6):1878, 1996.

233. Isliker, H., A. Anastasiadis, D. Vassiliadis and L. Vlahos. Solar flare cellular automata interpreted as discretized MHD equations. *Astronomy and Astrophysics*, 335(3):1085–92, 1998.

234. Jackson, E.A. *Perspectives of nonlinear dynamics - Volume 2*. Cambridge University Press, 1990.

235. Jaynes, E.T. Where do we stand on maximum entropy? In Levine, R.D. and M. Tribus, editor, *The Maximum Entropy Formalism*, pages 15–118. MIT Press, Cambridge, 1979.

236. Jeffreys, H. *Cartesian Tensors*. Cambridge University Press, Cambridge, 1965.

237. Jeffreys, H., and B. S. Jeffreys. *Methods of Mathematical Physics*. 3. Edition, Cambridge University Press, Cambridge, 1956.

238. Junk, M. Kinetic schemes in the case of low Mach numbers. *J. Comput. Phys.*, 151(2):947–968, 1999.

239. Kaandorp, J.A., C.P. Lowe, D. Frenkel, and P.M.A. Sloot. Effect of nutrient diffusion and flow on coral morphology. *Physical Review Letters*, 77(11):2328–31, 1996.

240. Kadanoff, L.P., G.R. McNamara, and G. Zanetti. From automata to fluid flow: Comparison of simulation and theory. *Phys. Rev. A*, 40(8):4527–4541, 1989.

241. Kahan, W. *Gauss-Seidel methods of solving large systems of linear equations*. PhD thesis, University of Toronto, 1958.

242. Kandhai, D., A. Koponen, A.G. Hoekstra, M. Kataja, J. Timonen, and P.M.A. Sloot. Lattice-Boltzmann hydrodynamics on parallel systems. *Computer Physics Communications*, 111(1-3):14–26, 1998a.

243. Kandhai, D., A. Koponen, A.G. Hoekstra, M. Kataja, J. Timonen, and P.M.A. Sloot. Implementation aspects of 3D lattice-BGK: boundaries, accuracy, and a new fast relaxation method. *Journal of Computational Physics*, 150(2):482–501, 1999.

244. Kandhai, D., D.J.-E. Vidal, A.G. Hoekstra, H. Hoefsloot, P. Iedema, and P.M.A. Sloot. A comparison between lattice-Boltzmann and finite-element simulations of fluid flow in static mixer reactors. *International Journal of Modern Physics C*, 9(8):1123–1128, 1998b.

245. Kari, J. Reversibility of 2D cellular automata is undecidable. *Physica D*, 45:375–385, 1990.

246. Karlin, I.V., A. Ferrante, and H.C. Öttinger. Perfect entropy functions of the lattice Boltzmann method. *Europhys. Lett.*, 47(2):182–188, 1999.

247. Karlin, I.V., A.N. Gorban, S. Succi, and V. Boffi. Maximum entropy principle for lattice kinetic equations. *Phys. Rev. Lett.*, 81(1):6–9, 1998.

248. Karlin, I.V. and S. Succi. Equilibria for discrete kinetic equations. *Physical Review E*, 58(4):R4053–R4056, 1998.

249. Karlin, I.V., S. Succi, and S. Orszag. Lattice Boltzmann method for irregular grids. *Physical Review Letters*, 82(26):5245–5248, 1999.

250. Karolyi, A., J. Kertesz, S. Havlin, H.A. Makse, and H.E. Stanley. Filling a silo with a mixture of grains: friction-induced segregation. *Europhysics Letters*, 44(3):386–392, 1998.

251. Kato, Y., K. Kono, T. Seta, D. Martinez and S. Chen. Amadeus project and microscopic simulation of boiling two-phase flow by the lattice-Boltzmann method. *International Journal of Modern Physics C*, 8(4):843–858, 1997.

252. Kauffman, S. A. Leben am Rande des Chaos. *Spektrum der Wissenschaft*, 10:90–99, 1991.

253. Kauffman, S. A. and R. G. Smith. Adaptive automata based on Darwinian selection. *Physica D*, 22:68–82, 1986.

254. Kauffman, S.A. Emergent properties of random cellular automata. *Physica D*, 10:145–156, 1984.

255. Kendon, V.M., J.-C. Desplat, P. Bladon, and M.E. Cates. 3D spinodal decomposition in the inertial regime. *Physical Review Letters*, 83(3):576–579, 1999.

256. Kepler, J. *Mysterium Cosmographicum*. Tübingen, 1596.

257. Kepler, J. *Mysterium Cosmographicum - The Secret of the Universe*. translated by A.M. Duncan, Abaris Books, New York, 1981.

258. Kerner, B.S., P. Konhaeuser, and M. Schilke. Dipole-layer effect in dense traffic flow. *Physics Letters A*, 215(1/2):45–56, 1996.

259. Kim, J., J. Lee, and K.-C. Lee. Nonlinear corrections to Darcy's law for flows in porous media. *Sae Mulli*, 38 (special issue):S119–S122, 1998.

260. Kingdon, R.D., P. Schofield and L. White. A lattice Boltzmann model for the simulation of fluid flow. *J. Phys. A: Math. Gen.*, 25:3559–3566, 1992.

261. Klar, A. Relaxation scheme for a lattice-Boltzmann type discrete velocity model and numerical Navier-Stokes limit. *Journal of Computational Physics*, 148(2):416–432, 1999.

262. Knackstedt, M.A., M. Sahimi and D.Y.C. Chan. Cellular-automata calculation of frequency-dependent permeability of porous media. *Phys. Rev. E*, 47(4):2593–2597, 1993.

263. Koch, D.L., R.J. Hill and A.S. Sangani. Brinkman screening and the covariance of the fluid velocity in fixed beds. *Physics of Fluids*, 10(12):3035–3037, 1998.

264. Koelman, J.M.V.A. A simple lattice Boltzmann scheme for Navier-Stokes fluid flow. *Europhys. Lett.*, 15(6):603–607, 1991.

265. Kohring, G.A. Parallelization of short- and long-range cellular automata on scalar, vector, SIMD and MIMD machines. *Int. J. Mod. Phys.*, C 2:755–772, 1991.

266. Kohring, G.A. Calculation of the permeability of porous media using hydrodynamic cellular automata. *J. Stat. Phys.*, 63:411–418, 1991a.

267. Kohring, G.A. Effect of finite grain size on the simulation of fluid flow in porous media. *Journal de Physique II France*, 1:87–90, 1991b.

268. Kohring, G.A. Limitations of a finite mean free path for simulation of fluid flow in porous media. *Journal de Physique II France*, 1:593–597, 1991c.

269. Kohring, G.A. An efficient hydrodynamic cellular automata for simulating fluids with large viscosities. *J. Stat. Phys.*, 66(3/4):1177–1184, 1992a.

270. Kohring, G.A. Calculation of drag coefficients via hydrodynamic cellular automata. *Journal de Physique*, 2:265ff, 1992b.

271. Koponen, A., D. Kandhai, E. Hellen, M. Alava, A. Hoekstra, M. Kataja, K. Niskanen, P. Sloot and J. Timonen. Permeability of three-dimensional random fiber webs. *Physical Review Letters*, 80(4):716–719, 1998a.

272. Koponen, A., M. Kataja, and J. Timonen. Tortuous flow in porous media. *Physical Review E*, 54(1):406–410, 1996.

273. Koponen, A., M. Kataja, and J. Timonen. Permeability and effective porosity of porous media. *Physical Review E*, 56(3):3319–3325, 1997.

274. Koponen, A., M. Kataja, J. Timonen, and D. Kandhai. Simulations of single-fluid flow in porous media. *International Journal of Modern Physics C*, 9(8):1505–1521, 1998b.

275. Kornreich, P.J. and J. Scalo. Supersonic lattice gases: restoration of Galilean invariance by nonlinear resonance effects. *Physica D*, 69(3/4):333–344, 1993.

276. Kougias, C. F. Numerical simulations of small-scale oceanic fronts of river discharge type with the lattice gas automata method. *J. Geophys. Res.*, 98(C10):18243–18255, 1993.

277. Krafczyk, M., M. Cerrolaza, M. Schulz and E. Rank. Analysis of 3D transient blood flow passing through an artificial aortic valve by Lattice-Boltzmann methods. *Journal of Biomechanics*, 31(5):453–462, 1998a.

278. Krafczyk, M., M. Schulz and E. Rank. Lattice-gas simulations of two-phase flow in porous media. *Communications in Numerical Methods in Engineering*, 14(8):709–717, 1998.

279. Krasheninnikov, S.I. and P.J. Catto. Lattice Boltzmann representations of neutral gas hydrodynamics. *Contributions to Plasma Physics*, 38(1-2):367–372, 1998.

280. Kullback, S. *Information Theory and Statistics.* Wiley, New York (reprint by Dover, 1968), 1959.

281. Kutrib, M., R. Vollmar and T. Worsch. Introduction to the special issue on cellular automata. *Parallel Computing*, 23(11):1567–1576, 1997.

282. Ladd, A. Short-time motion of colloidal particles: numerical simulation via a fluctuating lattice-Boltzmann equation. *Phys. Rev. Lett.*, 70(9):1339–1342, 1993.

283. Ladd, A. and D. Frenkel. Dissipative hydrodynamic interactions via lattice-gas cellular automata. *Physics of Fluids A*, 2(11):1921–1924, 1990.

284. Ladd, A.J.C. Numerical simulations of particulate suspensions via a discretized Boltzmann equation. Part 1. Theoretical foundation. *J. Fluid Mech.*, 271:285–310, 1994a.

285. Ladd, A.J.C. Numerical simulations of particulate suspensions via a discretized Boltzmann equation. Part 2. Numerical results. *J. Fluid Mech.*, 271:311–340, 1994b.

286. Lahaie, F. and J.R. Grasso. A fluid-rock interaction cellular automaton of volcano mechanics: application to the Piton de la Fournaise. *Journal of Geophysical Research*, 103(B5):963796–49, 1998.

287. Lamura, A., G. Gonnella and J.M. Yeomans. Modeling the dynamics of amphiphilic fluids. *International Journal of Modern Physics C*, 9(8):1469–1478, 1998.

288. Lamura, A., G. Gonnella and J.M. Yeomans. A lattice Boltzmann model of ternary fluid mixtures. *Europhysics Letters*, 45(3):314–320, 1999.

289. Landau, L.D. and E.M. Lifshitz. *Fluid Mechanics.* Pergamon Press and Addison-Wesley, 1959.

290. Lavallée, P., J. P. Boon and A. Noullez. Lattice Boltzmann equation for laminar boundary flow. *Complex Systems*, 3:317–330, 1989.

291. Ledermann, W., and S. Vajda (Edit.). *Handbook of Applicable Mathematics, Volume V: Combinatorics and Geometrie, Part A.* John Wiley & Sons, Chichester, 1985a.

292. Ledermann, W., and S. Vajda (Edit.). *Handbook of Applicable Mathematics, Volume V: Combinatorics and Geometrie, Part B.* John Wiley & Sons, Chichester, 1985b.

293. Levine, R.D. and M. Tribus, editor. *The Maximum Entropy Formalism.* MIT Press, Cambridge, 1979.

294. Lindgren, K., C. Moore, and M. Nordahl. Complexity of two-dimensional patterns. *J. Stat. Phys.*, 91(5-6):909–951, 1998.

295. Logan, J.D. *An Introduction to Nonlinear Partial Differential Equations.* John Wiley & Sons, 1994.

296. Luo, L.-S. Analytic solutions of linearized lattice Boltzmann equation for simple fluids. *J. Stat. Phys.*, 88(3/4):913–926, 1997a.

297. Luo, L.-S. Symmetry breaking of flow in 2D symmetric channels: simulations by lattice-Boltzmann method. *International Journal of Modern Physics C*, 8(4):859–867, 1997b.

298. Luo, L.-S. Unified theory of lattice Boltzmann models for nonideal gases. *Physical Review Letters*, 81(8):1618–1621, 1998.

299. Ma, S.-K. *Statistical Mechanics.* First Reprint, World Scientific, Philadelphia, 1993.

300. Machenhauer, B. The spectral method. In Global Atmospheric Research Programme (GARP), editor, *Numerical Methods Used in Atmospheric Models, Volume II*, pages 124–275. GARP Publications Series No. 17, 1979.

301. Maier, R.S. and R.S. Bernard. Accuracy of the lattice-Boltzmann method. *International Journal of Modern Physics C*, 8(4):747–752, 1997.

302. Maier, R.S., D.M. Kroll, H.T. Davis and R.S. Bernard. Pore-scale flow and dispersion. *International Journal of Modern Physics C*, 9(8):1523–1533, 1998a.

303. Maier, R.S., D.M. Kroll, Y.E. Kutsovsky, H.T. Davis and R.S. Bernard. Simulation of flow through bead packs using the lattice Boltzmann method. *Physics of Fluids*, 10(1):60–74, 1998b.

304. Maier, R.S., R.S. Bernard, and D.W. Grunau. Boundary conditions for the lattice Boltzmann method. *Physics of Fluids*, 8(7):1788–1801, 1996.

305. Makowiec, D. A note on the rule classification for square homogeneous cellular automata. *Physica A*, 236(3/4):353–362, 1997.

306. Malarz, K., K. Kulakowski, M. Antoniuk, M. Grodecki, and D. Stauffer. Some new facts of life. *International Journal of Modern Physics C*, 9(3):449–458, 1998.

307. Manna, S.S. and D.V. Khakhar. Internal avalanches in a granular medium. *Physical Review E*, 58(6):R6935–R6938, 1998.

308. Mareschal, M. and E. Kestemont. Experimental evidence for convective rolls in finite two-dimensional molecular models. *Nature*, 329:427–429, 1987.

309. Markus, M. and B. Hess. Isotropic cellular automaton for modeling excitable media. *Nature*, 347:56–58, 1990.

310. Martin, O., A. M. Odlyzko, and S. Wolfram. Algebraic properties of cellular automata. *Commun. Math. Phys.*, 93:219–258, 1984.

311. Martínez, D., S.Y. Chen and W.H. Matthaeus. Lattice Boltzmann magneto-hydrodynamics. *Phys. Plasmas*, 1(6):1850–1867, 1994.

312. Martínez, D.O., W.H. Matthaeus, S. Chen and D.C. Montgomery. Comparison of spectral method and lattice Boltzmann simulations of two-dimensional hydrodynamics. *Physics of Fluids*, 6(3):1285–1298, 1994.

313. Martys, N.S. and Hudong Chen. Simulation of multicomponent fluids in complex three-dimensional geometries by the lattice Boltzmann method. *Physical Review E*, 53(1):743–50, 1996.

314. Massaioli, F., R. Benzi and S. Succi. Exponential tails in two-dimensional Rayleigh-Bénard convection. *Europhys. Lett.*, 21:305–310, 1993.

315. Masselot, A. and B. Chopard. A lattice Boltzmann model for particle transport and deposition. *Europhysics Letters*, 42(3):259–264, 1998a.

316. Masselot, A. and B. Chopard. A multiparticle lattice-gas model for hydrodynamics. *International Journal of Modern Physics C*, 9(8):1221–1230, 1998b.

317. Matsukama, Y. Numerical simulation of complex flows by lattice gas automata method. *Journal of the Japan Society for Simulation Technology*, 17(3):220–228, 1998.

318. Matsukuma, Y., R. Takahashi, Y. Abe and H. Adachi. Lattice gas automata simulations of flow through porous media. In Sugimoto, J., editor, *Proceedings of the Workshop on Severe Accident Research (JAERI-Conf98-009)*, pages 128–133. Ibaraki-ken, Japan: Japan Atomic Energy Res. Inst., 1998.

319. Mayda, A. *Compressible Fluid Flow and Systems of Conservation Laws in Several Space Variables*. Springer, 1984.

320. McNamara, G. and B. Alder. Lattice Boltzmann simulation of high Reynolds number fluid flow in two dimensions. In Mareschal, M. and B.L. Holian, editor, *Microscopic Simulations of Complex Hydrodynamic Phenomena*, pages 125–136. Plenum Press, New York, 1992.

321. McNamara, G. and B. Alder. Analysis of the lattice Boltzmann treatment of hydrodynamics. *Physica A*, 194:218–228, 1993.

322. McNamara, G. and G. Zanetti. Use of the Boltzmann equation to simulate lattice-gas automata. *Phys. Rev. Lett.*, 61:2332–2335, 1988.

323. McNamara, G.R. Diffusion in a lattice gas automata. *Europhys. Lett.*, 12(4):329–334, 1990.

324. McNamara, G.R., A.L. Garcia and B.J. Alder. Stabilization of thermal lattice Boltzmann models. *J. Stat. Phys.*, 81(1/2):395–408, 1995.

325. McNamara, G.R., A.L. Garcia and B.J. Alder. A hydrodynamically correct thermal lattice Boltzmann model. *J. Stat. Phys.*, 87(5/6):1111–1121, 1997.

326. Mei, R. and W. Shyy. On the finite difference-based lattice Boltzmann method in curvilinear coordinates. *Journal of Computational Physics*, 143(2):426–448, 1998.

327. Meyer, D.A. Quantum mechanics of lattice gas automata: One-particle plane waves and potentials. *Physical Review E*, 55(5A):5261–5269, 1997a.

328. Meyer, D.A. Quantum lattice gases and their invariants. *International Journal of Modern Physics C*, 8(4):717–735, 1997b.

329. Meyer, D.A. Quantum mechanics of lattice gas automata: boundary conditions and other inhomogeneities. *Journal of Physics A*, 31(10):2321–2340, 1998.

330. Mielke, A. and R.B. Pandey. A computer simulation study of cell population in a fuzzy interaction model for mutating HIV. *Physica A*, 251(3-4):430–438, 1998.

331. Miller, W. Flow in the driven cavity calculated by the lattice Boltzmann method. *Physical Review E*, 51(4):3659–3669, 1995.

332. Miller, W. and K. Böttcher. Numerical study of flow and temperature patterns during the growth of $GaPO_4$ crystals using the lattice-Boltzmann. *International Journal of Modern Physics C*, 9(8):1567–1576, 1998.

333. Monaco, R., editor. *Discrete Kinetic Theory, Lattice Gas Dynamics and Foundations of Hydrodynamics*. World Scientific, Singapore, 1989.

334. Montgomery, D. and G. D. Doolen. Two cellular automata for plasma computations. *Complex Systems*, 1:830–838, 1987.

335. Moore, E.F. Machine models of self-reproduction. *Proc. Symp. Appl. Math.*, 14:17, 1962.

336. Morton, K.W. and D.F. Mayers. *Numerical Solution of Partial Differential Equations*. Cambridge University Press, 1994.

337. Munk, W.H. On the wind-driven ocean circulation. *J. Meteorol.*, 7(2):79–93, 1950.

338. Nagel, K. and M. Schreckenberg. A cellular automaton model for freeway traffic. *J. Phys. I France*, 2:2221–2229, 1992.

339. Nagel, K., D.E. Wolf, P. Wagner and P. Simon. Two-lane traffic rules for cellular automata: A systematic approach. *Physical Review E*, 58(2):1425–1437, 1998.

340. Nannelli, F. and S. Succi. The lattice Boltzmann equation on irregular lattices. *J. Stat. Phys.*, 68(3/4):401–407, 1992.

341. Nasilowski, R. An arbitrary-dimensional cellular-automaton fluid model with simple rules. Proceedings in Dissipative Structrures in Transport Processes and Combustion (Interdisciplinary Seminar, Bielefeld 1989), Springer-Verlag, Heidelberg, 1990.

342. Nasilowski, R. A cellular-automaton fluid model with simple rules in arbitrary many dimensions. *J. Stat. Phys.*, 65(1/2):97–138, 1991.

343. Nicodemi, M. A phenomenological theory of dynamic processes in granular media. *Physica A*, 257(1-4):448–453, 1998.

344. Nie, X., Y.-H. Qian, G.D. Doolen and S. Chen. Lattice Boltzmann simulation of the two-dimensional Rayleigh-Taylor instability. *Physical Review E*, 58(5):6861–6864, 1998.

345. Niimura, H. Deformable porous structure of fluids by multi-fluid lattice-gas automaton. *Physics Letters A*, 245(5):366–372, 1998.

346. Nishinari, K. and D. Takahashi. Analytical properties of ultradiscrete Burgers equation and rule-184 cellular automaton. *Journal of Physics A (Mathematical and General)*, 31(24):5439–5450, 1998.

347. Nobel, D.R., S. Chen, J.C. Georgiadis and R. Buckius. A consistent hydrodynamic boundary condition for the lattice Boltzmann method. *Phys. Fluids*, 7(1):203–209, 1995.

348. Noble, D.R. and J.R. Torczynski. A lattice-Boltzmann method for partially saturated computational cells. *International Journal of Modern Physics C*, 9(8):1189–1201, 1998.

349. Noble, D.R.; Georgiadis, J.G.; Buckius, R.O. Comparison of accuracy and performance for lattice Boltzmann and finite difference simulations of steady viscous flow. *International Journal for Numerical Methods in Fluids*, 23(1):1–18, 1996.

350. Noble, D.R., J.G. Georgiadis and R.O. Buckins. Direct assessment of lattice Boltzmann hydrodynamics and boundary conditions for recirculating flows. *J. Stat. Phys.*, 81(1/2):17–34, 1995.

351. Nordfalk, J. and P. Alstrom. Phase transitions near the "game of Life". *Physical Review E*, 54(2):R1025–8, 1996.

352. Ohashi, H., Y.Chen, and M. Akiyama. Simulation of shock-interface interaction using a lattice Boltzmann model. *Nuclear Engineering and Design*, 155(1/2):67–71, 1995.

353. Olson, J.F. and D.H. Rothman. Three-dimensional immiscible lattice gas: application to sheared phase separation. *J. Stat. Phys.*, 81(1/2):199–222, 1995.

354. Olson, J.F. and D.H. Rothman. Two-fluid flow on sedimentary rock: simulation, transport and complexity. *J. Fluid Mech.*, 341:343–370, 1997.

355. Orlandini, E., M.R. Swift and J.M. Yeomans. A lattice Boltzmann model of binary-fluid mixtures. *Europhys. Lett.*, 32(6):463–468, 1995.

356. Orlanski, I. A simple boundary condition for unbounded hyperbolic flows. *J. Comput. Phys.*, 21:251–269, 1976.

357. Orszag, S.A. Numerical simulation of incompressible flow within simple boundaries: Accuracy. *J. Fluid Mech.*, 49(1):75–112, 1971.

358. Orszag, S.A., Y.H. Qian, and S. Succi. Applications of lattice Boltzmann methods to fluid dynamics. In , editor, *Progress and Challenges in CFD Methods and Algorithms (AGARD-CP-578)*, pages 25/1–10. Neuilly sur Seine, France: AGARD, 1996.

359. O'Toole, D.V., P.A. Robinson, and M.R. Myerscough. Self-organized criticality in termite architecture: a role for crowding in ensuring ordered nest expansion. *J. theor. Biol.*, 198:305–327, 1999.

360. Pacanowski, R.C., K.W. Dixon, and A. Rosati. The GFDL Modular Ocean Model Users Guide Version 1.0. GFDL Ocean Group Tech. Rep. No. 2, 46pp, 1991.

361. Pavlo, P., G. Vahala and L. Vahala. Higher order isotropic velocity grids in lattice methods. *Physical Review Letters*, 80(18):3960–3963, 1998a.

362. Pavlo, P., G. Vahala, L. Vahala and M. Soe. Linear stability analysis of thermo-lattice Boltzmann models. *Journal of Computational Physics*, 139(1):79–91, 1998b.

363. Peitgen, H.-O., A. Rodenhausen, and G. Skordev. Self-similar functions generated by cellular automata. *Fractals*, 6(4):371–394, 1998.

364. Peitgen, H.-O., H. Jürgens and D. Saupe. *Chaos and Fractals*. Springer, New York, 1992.

365. Peng, G. and T. Ohta. Velocity and density profiles of granular flow in channels using a lattice gas automaton. *Physical Review E*, 55(6 A):6811–6820, 1997.

366. Peng, G., H. Xi, and C. Duncan. Lattice Boltzmann method on irregular meshes. *Physical Review E*, 58(4):R4124–R4127, 1998.

367. Peng, G., H. Xi, and C. Duncan. Finite volume scheme for the lattice Boltzmann method on unstructured meshes. *Physical Review E*, 59(4):4675–4682, 1999.

368. Penrose, R. The rôle of aesthetics in pure and applied mathematical research. *Bull. Inst. Math. and its Appl.*, 10:266–269, 1974.

369. Penrose, R. Pentaplexity - A class of non-periodic tilings of the plane. *Mathematical Intelligencer*, 2:32–37, 1979.

370. Phillips, N. A. The general circulation of the atmosphere: a numerical experiment. *Quat. J. Roy. Meteor. Soc.*, 82:123–164, 1956.

371. Phillips, N. A. An example of non-linear computational instability. The Atmosphere and the Sea in Motion, Rossby Memorial Volume, New York, Rockefeller Institute Press, 501-504, 1959.

372. Press, W.H., B.P. Flannery, S.A. Teukolsky, and W.T. Vetterling. *Numerical Recipes in C*. Cambridge University Press, Cambridge, 1992.

373. Press, W.H., B.P. Flannery, S.A. Teukolsky, and W.T. Vetterling. *Numerical Recipes in FORTRAN*. Cambridge University Press, Cambridge, 1992b.

374. Pulsifer, J.E. and C.A. Reiter. One tub, eight blocks, twelve blinkers and other views of life. *Computers & Graphics*, 20(3):457–462, 1996.

375. Punzo, G., F. Massaioli, and S. Succi. High-resolution lattice-Boltzmann computing on the IBM SP1 scalable parallel computer. *Computers in Physics*, 8(6):705–711, 1994.

376. Qi, D. Non-spheric colloidal suspensions in three-dimensional space. *International Journal of Modern Physics C*, 8(4):985–997, 1997.

377. Qi, D. Lattice-Boltzmann simulations of particles in non-zero-Reynolds-number flows. *Journal of Fluid Mechanics*, 385:41–62, 1999.

378. Qian, Y.-H. Fractional propagation and the elimination of staggered invariants in lattice-BGK models. *International Journal of Modern Physics C*, 8(4):753–761, 1997.

379. Qian, Y.-H. and S. Chen. Finite size effect in lattice-BGK models. *International Journal of Modern Physics C*, 8(4):763–771, 1997.

380. Qian, Y.-H. and Y. Zhou. Complete Galilean-invariant lattice BGK models for the Navier-Stokes equation. *Europhysics Letters*, 42(4):359–364, 1998.

381. Qian, Y.H. Simulating thermohydrodynamics with lattice BGK models. *J. Sci. Comput.*, 8(3):231–242, 1993.

382. Qian, Y.H. and S.A. Orszag. Lattice BGK models for the Navier-Stokes equation: nonlinear deviation in compressible regimes. *Europhys. Lett.*, 21(3):255–259, 1993.

383. Qian, Y.H. and S.A. Orszag. Scalings in diffusion-driven reaction $A + B \rightarrow C$: Numerical simulations by lattice BGK models. *J. Stat. Phys.*, 81(1/2):237–254, 1995.

384. Qian, Y.H., D. d'Humières, and P. Lallemand. Lattice BGK models for Navier-Stokes equation. *Europhys. Lett.*, 17(6):479–484, 1992.

385. Qian, Y.H., S. Succi and S.A. Orszag. Recent advances in lattice Boltzmann computing. In Stauffer, D., editor, *Annual Review of Computational Physics III*, pages 195–242. World Scientific, Singapore, 1995.

386. Rakotomalala, N., D. Salin, and P. Watzky. Simulations of viscous flows of complex fluids with a Bhatnagar, Gross, and Krook lattice gas. *Physics of Fluids*, 8(11):3200–2, 1996.

387. Rapaport, D.C. *The Art of Molecular Dynamics Simulation*. Cambridge University Press, 1995.

388. Reider, M.B. and J.D. Sterling. Accuracy of discrete-velocity BGK models for the simulation of the incompressible Navier-Stokes equations. *Comput. Fluids*, 24:459–467, 1995.

389. Rem, P. C. and J. A. Somers. Cellular automata algorithms on a transputer network. In R. Monaco, editor, *Discrete Kinematic Theory, Lattice Gas Dynamics and Foundations of Hydrodynamics*, pages 268–275. World Scientific, Singapore, 1989.

390. Renwei, M. and S. Wei. On the finite difference-based lattice Boltzmann method in curvilinear coordinates. *J. Comput. Phys.*, 143(2):426–448, 1998.

391. Rényi, A. *Probability Theory.* North Holland, 1970.

392. Reynolds, O. An experimental investigation of the circumstances which determine whether the motion of water shall be direct or sinuous, and of the law of resistance in parallel channels. *Phil. Trans. Roy. Soc.*, 174:935–982, 1883.

393. Richter, S. and R.F. Werner. Ergodicity of quantum cellular automata. *J. Stat. Phys.*, 82(3/4):963–98, 1996.

394. Rickert, M., K. Nagel, M. Schreckenberg, and A. Latour. Two lane traffic simulations using cellular automata. *Physica A*, 231(4):534–550, 1996.

395. Rivet, J.-P. Green-Kubo formalism for lattice gas hydrodynamics and Monte-Carlo evaluation of shear viscosities. *Complex Systems*, 1:839–851, 1987.

396. Rivet, J.-P. *Hydrodynamique par la méthode des gaz sur réseaux.* PhD thesis, Université de Nice, 1988.

397. Rivet, J.-P. Brisure spontanée de symétrie dans le sillage tri-dimensionnel d'un cylindre allongé, simulé par la methode des gaz sur réseaux (Spantaneous symmetry-breaking in the 3-D wake of a long cylinder, simulated by the lattice gas method). *C. R. Acad. Sci. II*, 313:151–157, 1991.

398. Rivet, J.-P., M. Hénon, U. Frisch, and D. d'Humières. Simulating fully three-dimensional external flow by lattice gas methods. *Europhys. Lett.*, 7:231–236, 1988.

399. Rivet, Jean-Pierre and Jean Pierre Boon. *Lattice Gas Hydrodynamics.* Cambridge University Press, (to be published).

400. Røed, P. and O.M. Smedstad. Open boundary conditions for forced waves in a rotating fluid. *SIAM J. Sci. Stat. Comput.*, 5(2):414–426, 1984.

401. Rood, R. B. Numerical advection algorithms and their role in atmospheric transport and chemistry models. *Rev. Geophys.*, 25(1):71–100, 1987.

402. Rothman, D. H. Cellular-automaton fluids: A model for flow in porous media. *Geophysics*, 53(4):509–518, 1988.

403. Rothman, D. H., and J. M. Keller. Immiscible Cellular-Automaton Fluids. *J. Stat. Phys.*, 52:1119–1127, 1988.

404. Rothman, D.H. and S. Zaleski. Lattice-gas models of phase separation: interfaces, phase transitions, and multiphase flow. *Rev. Modern Phys.*, 66(4):1417–1479, 1994.

405. Rothman, D.H. and S. Zaleski. *Lattice-Gas Cellular Automata - Simple Models of Complex Hydrodynamics.* Cambridge University Press, 1997.

406. Ruján, P. Cellular automata and statistical mechanical models. *J. Stat. Phys.*, 49:139–222, 1987.

407. Sahimi, M. Flow phenomena in rocks: from continuum models to fractals, percolation, cellular automata, and simulated annealing. *Rev. Mod. Phys.*, 65(4):1393–1534, 1993.

408. Salmon, R. The lattice Boltzmann method as a basis for ocean circulation modeling. *J. Mar. Res.*, 57:503–535, 1999.

409. Schadschneider, A. and M. Schreckenberg. Garden of Eden states in traffic models. *Journal of Physics A*, 31(11):L225–L231, 1998.

410. Schelkle, M., M. Rieber und A. Frohn. Numerische Simulation von Tropfenkollisionen. *Spektrum der Wissenschaft*, 1:72–79, 1999.

411. Schrandt, R. and S. Ulam. On recursively defined geometrical objects and patterns of growth. In Burks, A. W., editor, *Essays on Cellular Automata*, pages 232–243. University of Illinois Press, Urbana, 1970.

412. Schreckenberg, M., A. Schadschneider, K. Nagel, and N. Ito. Discrete stochastic models for traffic flow. *Physical Review E*, 51(4):2939–49, 1995.

413. Sehgal, B.R., R.R. Nourgaliev, and T.N. Dinh. Numerical simulation of droplet deformation and break-up by lattice-Boltzmann method. *Progress in Nuclear Energy*, 34(4):471–488, 1999.

414. Semtner, A.J. Introduction to "A numerical method for the study of the circulation of the world ocean". *J. Comput. Phys.*, 135:149–153, 1997.

415. Shan, X. and G. Doolen. Diffusion in a multi-component lattice Boltzmann equation model. *Physical Review E*, 54(4A):3614–20, 1996.

416. Shan, X. and H. Chen. Lattice Boltzmann model for simulating flows with multiple phases and components. *Physical Review E*, 47(3):1815–9, 1993.

417. Shan, X. and H. Chen. Simulation of nonideal gases and liquid-gas phase transitions by the lattice Boltzmann equation. *Physical Review E*, 49(4):2941–8, 1994.

418. Shannon, C. E. A mathematical theory of communications. *Bell System Tech. J.*, 27:379,623, 1948.

419. Shannon, C.E. and W. Weaver. *The Mathematical Theory of Communication*. University of Illinois Press, Urbana, 1949.

420. Shimomura, T., G. Doolen, B. Hasslacher, and C. Fu. Calculations using lattice gas techniques. *Los Alamos Science*, 15:201–210, 1988.

421. Sigmund, K. *Games of Life: Explorations in Ecology, Evolution, and Behavior.* Oxford University Press, 1993.

422. Signorini, J. Complex computing with cellular automata. In Manneville, P., N. Boccara, G. Y. Vichniac and R. Bidaux, editor, *Cellular Automata and Modeling of Complex Physical Systems*, pages 57–72. Springer, Berlin, 1989.

423. Simons, N.R.S., G.E. Bridges, and M. Cuhaci. A lattice gas automaton capable of modeling three-dimensional electromagnetic fields. *J. Comput. Phys.*, 151(2):816–835, 1999.

424. Siregar, P., J.P. Sinteff, M. Chahine and P. Lebeux. A cellular automata model of the heart and its coupling with a qualitative model. *Computers and Biomedical Research*, 29(3):222–46, 1996.

425. Siregar, P., J.P. Sinteff, N. Julen and P. Le Beux. An interactive 3D anisotropic cellular automata model of the heart. *Computers and Biomedical Research*, 31(5):323–347, 1998.

426. Skordos, P. A. Initial and boundary conditions for the lattice Boltzmann method. *Phy. Rev. E*, 48(6):4823–4842, 1993.

427. Slone, D.M. and G.H. Rodrigue. Efficient biased random bit generation for parallel lattice gas simulations. *Parallel Computing*, 22(12):1597–1620, 1997.

428. Sofonea, V. Lattice Boltzmann approach to collective-particle interactions in magnetic fluids. *Europhys. Lett.*, 25(5):385–390, 1994.

429. Sofonea, V. Two-phase fluid subjected to terrestrial or space conditions: a lattice Boltzmann study. *International Journal of Modern Physics C*, 7(5):695–704, 1996.

430. Somers, J. A. and P. C. Rem. The construction of efficient collision tables for fluid flow. In Manneville, P., N. Boccara, G. Y. Vichniac and R. Bidaux, editor, *Cellular Automata and Modeling of Complex Physical Systems*, pages 161–177. Springer, Berlin, 1989.

431. Spaid, M.A.A. and F.R. Phelan, Jr. Lattice Boltzmann methods for modeling microscale flow in fibrous porous media. *Physics of Fluids*, 9(9):2468–2474, 1997.

432. Spaid, M.A.A. and F.R. Phelan, Jr. Modeling void formation dynamics in fibrous porous media with the lattice Boltzmann method. *Composites Part A (Applied Science and Manufacturing)*, 29A(7):749–755, 1998.

433. Spall, M.A. and W.R. Holland. A nested primitive equation model for oceanic applications. *J. Phys. Oceanogr.*, 21:205–220, 1991.

434. Stauffer, D. Classification of square lattice cellular automata. *Physica A*, 157:645–655, 1989.

435. Sterling, J.D. and S. Chen. Stability analysis of lattice Boltzmann methods. *J. Computational Physics*, 123:196–206, 1996.

436. Stevens, D.P. The open boundary conditions in the United Kingdom Fine-Resolution Antarctic Model. *J. Phys. Oceanogr.*, 21:1494–1499, 1991.

437. Stockman, H.W., C. Li and J.L. Wilson. A lattice-gas and lattice Boltzmann study of mixing at continuous fracture junctions: importance of boundary conditions. *Geophysical Research Letters*, 24(12):1515–1518, 1997.

438. Stockman, H.W., C.T. Stockman, and C.R. Carrigan. Modeling viscous segregation in immiscible fluids using lattice-gas automata. *Nature*, 34:523ff, 1990.

439. Stockman, H.W., R.J. Glass, C. Cooper, and H. Rajaram. Accuracy and computational efficiency in 3D dispersion via lattice-Boltzmann: models for dispersion in rough fractures and double-diffusive fingering. *International Journal of Modern Physics C*, 9(8), 1998.

440. Stokes, G.C. On the effect of the internal friction of fluids on the motion of pendulums. *Trans. Camb. Phil. Soc.*, 9 (II):8–14, 1851.

441. Stommel, H. The westward intensification of wind-driven currents. *Trans. Am. Geophys. Union*, 29:202–206, 1948.

442. Stueckelberg, E. C. G. Théorème H et unitarité de S. *Helv. Phys. Acta*, 25:577–580, 1952.

443. Stumpf, H. und A. Rieckers. *Thermodynamik, 2 Bände*. Vieweg, Braunschweig, 1976.

444. Suárez, A. and J.P. Boon. Nonlinear hydrodynamics of lattice-gas automata with semi-detailed balance. *International Journal of Modern Physics C*, 8(4):653–674, 1997.

445. Succi, S. Numerical solution of the Schrödinger equation using discrete kinetic theory. *Physical Review E*, 53(2):1969–1975, 1996.

446. Succi, S. Lattice quantum mechanics: an application to Bose-Einstein condensation. *International Journal of Modern Physics C*, 9(8):1577–85, 1998.

447. Succi, S. Unified lattice Boltzmann schemes for turbulence and combustion. *Zeitschrift fur Angewandte Mathematik und Mechanik*, 78, suppl.(1):S129–132, 1998.

448. Succi, S. and F. Nannelli. The finite volume formulation of the lattice Boltzmann equation. *Transport Theory Stat. Phys.*, 23:163–171, 1994.

449. Succi, S. and P. Vergari. A lattice Boltzmann scheme for semiconductor dynamics. *VLSI Design*, 6(1-4):137–140, 1998..

450. Succi, S. and R. Benzi. Lattice Boltzmann equation for quantum mechanics. *Physica D*, 69(3/4):327–332, 1993.

451. Succi, S., G. Amati and R. Benzi. Challenges in lattice Boltzmann computing. *J. Stat. Phys.*, 81(1/2):5–16, 1995.

452. Succi, S., G. Bella and F. Papetti. Lattice kinetic theory for numerical combustion. *Journal of Scientific Computing*, 12(4):395–408, 1997.

453. Succi, S., M. Vergassola and R. Benzi. Lattice Boltzmann scheme for two-dimensional magnetohydrodynamics. *Phys. Rev. A*, 43(8):4521–4524, 1991.

454. Sun, Chenghai. Multispecies lattice Boltzmann models for mass diffusion. *Acta Mechanica Sinica*, 30(1):20–26, 1998.

455. Sverdrup, H.U. Wind-driven currents in a baroclinic ocean, with application to the equatorial currents of the eastern Pacific. *Proc. Natl. Acad. Sci. US.*, 33:319–326, 1947.

456. Swift, M.R., S.E. Orlandini, W.R. Osborn, and J.M. Yeomans. Lattice Boltzmann simulations of liquid-gas and binary fluid systems. *Physical Review E*, 54(5):5041–52, 1996.

457. Takada, N. and M. Tsutahara. Evolution of viscous flow around a suddenly rotating circular cylinder in the lattice Boltzmann method. *Computers & Fluids*, 27(7):807–828, 1998.

458. Takalo, J., J. Timonen, A. Klimas, J. Vaklivia, and D. Vassiliadis. Nonlinear energy dissipation in a cellular automaton magnetotail field model. *Geophysical Research Letters*, 26(13):1813–1816, 1999.

459. Takesue, S. Reversible cellular automata and statistical mechanics. *Physical Review Letters*, 59(22):2499–2502, 1987.

460. Takesue, S. Ergodic properties and thermodynamic behavior of elementary reversible cellular automata. I. Basic properties. *J. Stat. Phys.*, 56:371–402, 1989.

461. Takesue, S. Relaxation properties of elementary reversible cellular automata. *Physica D*, 45:278–284, 1990.

462. Takesue, S. Staggered invariants in cellular automata. *Complex Systems*, 9(2):149–168, 1995.

463. Tan, M.-L., Y.H. Qian, I. Goldhirsch and S.A. Orszag. Lattice-BGK approach to simulating granular flows. *J. Stat. Phys.*, 81(1/2):87–104, 1995.

464. Teixeira, C.M. Digital Physics simulation of lid-driven cavity flow. *International Journal of Modern Physics C*, 8(4):685–696, 1997.

465. Teixeira, C.M. Incorporating turbulence models into the lattice-Boltzmann method. *International Journal of Modern Physics C*, 9(8):1159–1175, 1998.

466. Teixeira, J. Stable schemes for partial differential equations: the one-dimensional diffusion equation. *J. Comput. Phys.*, 153:403–417, 1999.

467. Theissen, O., G. Gompper, and D.M. Kroll. Lattice-Boltzmann model of amphiphilic systems. *Europhysics Letters*, 42(4):419–424, 1998.

468. Toffoli, T. and N. Margolus. Invertible cellular automata: A review. *Physica D*, 45:229–253, 1990.

469. Tölke, J., M. Krafczyk, M. Schulz, E. Rank, and R. Berrios. Implicit discretization and nonuniform mesh refinement approaches for FD discretizations of LBGK models. *International Journal of Modern Physics C*, 9(8):1143–1157, 1998.

470. Tougaw, P.D. and C.S. Lent. Dynamic behavior of quantum cellular automata. *Journal of Applied Physics*, 80(8):4722–4736, 1996.

471. Tribel, O. and J.P. Boon. Entropy and correlations in lattice-gas automata without detailed balance. *International Journal of Modern Physics C*, 8(4):641–652, 1997.

472. Tsujimoto, S. and R. Hirota. Ultradiscrete KdV equation. *Journal of the Physical Society of Japan*, 67(6):1809–1810, 1998.

473. Tsumaya, A. and H. Ohashi. Immiscible lattice gas with long-range interaction. *International Journal of Modern Physics C*, 8(4):697–703, 1997.

474. Turing, A.M. On computable numbers, with an application to the Entscheidungsproblem. *Proc. Lond. Math. Soc.*, 42:230–265, 1936-37.

475. Twining, C. J. The limiting behavior of non-cylindrical elementary cellular automata. *Complex Systems*, 6(5):417–432, 1992.

476. Uhlenbeck, G. and G. Ford. *Lectures in Statistical Mechanics.* Providence, 1963.

477. Ujita, H., S. Nagata, M. Akiyama, and M. Naitoh. Development of LGA and LBE 2D parallel programs. *International Journal of Modern Physics C*, 9(8):1203–1220, 1998.

478. Ulam, S. Random processes and transformations. Proceedings of the International Congress on Mathematics, 1950, Vol. 2, p. 264-275, 1952.

479. Ulam, S. On some mathematical problems connected with patterns of growth of figures. Proceedings of Symposia in Applied Mathematics 14, American Mathematical Society, Providence, p. 215-224, 1962.

480. Vahala, G., P. Pavlo and L. Vahala. Lattice Boltzmann representation of neutral turbulence in the cold gas blanket divertor regime. *Czechoslovak Journal of Physics*, 48(8):953–962, 1998a.

481. Vahala, G., P. Pavlo, L. Vahala and M. Soe. Determination of eddy transport coefficients in thermo-lattice Boltzmann modeling of two-dimensional turbulence. *Czechoslovak Journal of Physics*, 46(11):1063–1083, 1996.

482. Vahala, G., P. Pavlo, L. Vahala and N.S. Martys. hermal lattice-Boltzmann models (TLBM) for compressible flows. *International Journal of Modern Physics C*, 9(8):1247–1261, 1998b.

483. van Coevorden, D. V., M. H. Ernst, R. Brito and J. A. Somers. Relaxation and transport in FCHC lattice gases. *J. Stat. Phys.*, 74(5/6):1085–1115, 1994.

484. van der Hoef, M.A. and D. Frenkel. Long-time tails of the velocity autocorrelation function in two- and three-dimensional lattice-gas cellular automata: A test of mode-coupling theory. *Phys. Rev. A*, 41:4277–4284, 1990.

485. van der Hoef, M.A., M. Dijkstra and D. Frenkel. Velocity autocorrelation function in a four-dimensional lattice gas. *Europhys. Lett.*, 17(1):39–43, 1992.

486. van der Sman, R.G.M. Lattice-Boltzmann scheme for natural convection in porous media. *International Journal of Modern Physics C*, 8(4):879–888, 1997.

487. van der Sman, R.G.M. and M.H. Ernst. Diffusion lattice Boltzmann scheme on a orthorhombic lattice. *J. Stat. Phys.*, 94(1-2):203–217, 1999.

488. van Dyke, Milton. *An Album of Fluid Motion.* Parabolic Press, Stanford, CA, 1982.

489. van Genabeek, O. and D.H. Rothman. Macroscopic manifestations of microscopic flows through porous media: phenomenology from simulation. *Ann. Rev. Earth Planet. Sci.*, 24:63–87, 1996.

490. van Genabeek, O. and D.H. Rothman. Critical behavior in flow through a rough-walled channel. *Physics Letters A*, 255(1-2):31–36, 1999.

491. Vanag, V.K. and G. Nicolis. Nonlinear chemical reactions in dispersed media: The effect of slow mass exchange on the steady-state of the Schlogl models. *Journal of Chemical Physics*, 110(9):4505–4513, 1999.

492. Vandewalle, N. and M. Ausloos. Evolution motivated computer models. In Stauffer, D., editor, *Annual Review of Computational Physics III*, pages 45–85. World Scientific, Singapore, 1995.

493. Verberg, R. and A.J.C. Ladd. Simulation of low-Reynolds-number flow via a time-independent lattice-Boltzmann method. *Physical Review E*, 60(3):3366–3373, 1999.

494. Vergassola, M., R. Benzi, and S. Succi. On the hydrodynamic behaviour of the lattice Boltzmann equation. *Europhys. Lett.*, 13:411–416, 1990.

495. Verheggen, T.M.M., editor. *Numerical methods for the simulation of multiphase and complex flow : proceedings of a workshop held at Koninklijke/Shell-Laboratorium, Amsterdam, the Netherlands, 30 May-1 June 1990*. Springer, Berlin, 1992.

496. Verlet, L. Computer "experiments" on classical fluids. I. Thermodynamical properties of Lennard-Jones molecules. *Phys. Rev.*, 159(1):98–103, 1967.

497. Vogeler, A., and D.A. Wolf-Gladrow. Pair interaction lattice gas simulations: flow past obstacles in two and three dimensions. *J. Stat. Phys.*, 71(1/2):163–190, 1993.

498. Vollmar, R., W. Erhard and V. Jossifov, editor. *Parcella '96. Proceedings of the VII. International Workshop on Parallel Processing by Cellular Automata and Arrays*. Akademie Verlag, Berlin, 1996.

499. von Neumann, J. *The Theory of Self-Reproducing Automata*. University of Illinois Press, Urbana, 1966.

500. Voorhees, B. Nearest neighbor cellular automata over Z_2 with periodic boundary conditions. *Physica D*, 45:26–35, 1990.

501. Voorhees, B. Predecessors of cellular automata states. II. Pre-images of finite sequences. *Physica D*, 73:136–151, 1994.

502. Voorhees, B. and S. Bradshaw. Predecessors of cellular automata states. III. Garden of Eden classification of cellular automata. *Physica D*, 73:152–167, 1994.

503. Voorhees, B.H. *Computational Analysis of One-Dimensional Cellular Automata*. World Scientific, 1996.

504. Wagner, A.J. An H-theorem for the lattice Boltzmann approach to hydrodynamics. *Europhysics Letters*, 44(2):144–149, 1998a.

505. Wagner, A.J. Spinodal decomposition in two-dimensional binary fluids. *International Journal of Modern Physics C*, 9(8):1373–1382, 1998b.

506. Wagner, A.J. and J.M. Yeomans. Effect of shear on droplets in a binary mixture. *International Journal of Modern Physics C*, 8(4):773–782, 1997.

507. Wagner, A.J. and J.M. Yeomans. Breakdown of scale invariance in the coarsening of phase-separating binary fluids. *Physical Review Letters*, 80(7):1429–1432, 1998.

508. Wagner, A.J. and J.M. Yeomans. Phase separation under shear in two-dimensional binary fluids. *Physical Review E*, 59(4):4366–4373, 1999.

509. Wagner, L. and F. Hayot. Lattice Boltzmann simulations of flow past a cylindrical obstacle. *J. Stat. Phys.*, 81(1/2):63–70, 1995.

510. Waite, M.E., G. Shemin, and H. Spetzler. A new conceptual model for fluid flow in discrete fractures: an experimental and numerical study. *Journal of Geophysical Research*, 104(B6):13049–13059, 1999.

511. Waite, M.E., G. Shemin, H. Spetzler and D.B. Bahr. The effect of surface geometry on fracture permeability: a case study using a sinusoidal fracture. *Geophysical Research Letters*, 25(6):813–816, 1998.

512. Wang, B.-H., Y.R. Kwong, and P.M. Hui. Statistical mechanical approach to cellular automaton models of highway traffic flow. *Physica A*, 254(1-2):122–134, 1998.

513. Warren, P.B. Electroviscous transport problems via lattice-Boltzmann. *International Journal of Modern Physics C*, 8(4):889–898, 1997.

514. Watrous, J. On one-dimensional quantum cellular automata. In , editor, *Proceedings. 36th Annual Symposium on Foundations of Computer Science (Cat. No.95CB35834)*, pages 528–537. Los Alamitos, CA, USA: IEEE Comput. Soc. Press, 1995.

515. Weig, F.W.J., P.V. Coveney and B.M. Boghosian. Lattice-gas simulations of minority-phase domain growth in binary immiscible and ternary amphiphilic fluids. *Physical Review E*, 56(6):6877–6888, 1997.

516. Weimar, J.R. Cellular automata for reaction-diffusion systems. *Parallel Computing*, 23(11):1699–1715, 1997.

517. Weimar, J.R. and J.P. Boon. Nonlinear reactions advected by a flow. *Physica A*, 224:207–215, 1996.

518. Weimar, J.R., D. Dab, J.P. Boon, and S. Succi. Fluctuation correlations in reaction-diffusion systems: reactive lattice gas automata approach. *Europhys. Lett.*, 20(7):627–632, 1992.

519. Welander, P. On the temperature jump in a rarefied gas. *Arkiv för Fysik*, 7(44):507–553, 1954.

520. Weyl, H. *Symmetry*. Princeton University Press, 1989.

521. Wiener, N. and A. Rosenblueth. The mathematical formulation of the problem of conduction of impulses in a network of connected excitable elements, specifically in cardiac muscle. *Archivos del Instituto de Cardiologia de Mexico*, 16:202–265, 1946.

522. Wolf-Gladrow, D.A. A lattice Boltzmann equation for diffusion. *J. Stat. Phys.*, 79(5/6):1023–1032, 1995.

523. Wolf-Gladrow, D.A. and A. Vogeler. Pair interaction lattice gas on general purpose computers: FORTRAN or C? *Int. J. Mod. Phys., C*, 3:1179–1187, 1992.

524. Wolf-Gladrow, D.A., R. Nasilowski, and A. Vogeler. Numerical simulations of fluid dynamics with a pair interaction automaton in two dimensions. *Complex Systems*, 5:89–100, 1991.

525. Wolfram, S. Statistical mechanics of cellular automata. *Rev. Mod. Phys.*, 55:601–644, 1983.

526. Wolfram, S. Computer software in science and mathematics. *Scientific American*, 251:188–203, 1984.

527. Wolfram, S. Cellular automata as models of complexity. *Nature*, 311:419–424, 1984a.

528. Wolfram, S. Universality and complexity in cellular automata. *Physica D*, 10:1–35, 1984b.

529. Wolfram, S. Twenty problems in the theory of cellular automata. *Physica Scripta*, T9:170–183, 1985.

530. Wolfram S. Cellular Automaton Fluids 1: Basic Theory. *J. Stat. Phys.*, 45(3/4):471–526, 1986.

531. Wolfram, S., editor. *Theory and Applications of Cellular Automata*. World Publishing Co., Singapore, 1986.

532. Wolfram, S. *Cellular Automata and Complexity*. Addison-Wesley, Reading, MA, 1994.

533. Worthing, R.A., J. Mozer and G. Seeley. Stability of lattice Boltzmann methods in hydrodynamic regimes. *Physical Review E*, 56(2):2243–2253, 1997.

534. Xi, H. and C. Duncan. Lattice Boltzmann simulations of three-dimensional single droplet deformation and breakup under simple shear flow. *Physical Review E*, 59(3):3022–3026, 1999.

535. Xi, H., G. Peng, and S.-H. Chou. Finite-volume lattice Boltzmann method. *Physical Review E*, 59(5):6202–6205, 1999a.

536. Xi, H., G. Peng, and S.-H. Chou. Finite-volume lattice Boltzmann schemes in two and three dimensions. *Physical Review E*, 60(3):3380–3388, 1999b.

537. Xu, K. and L.-S. Luo. Connection between lattice-Boltzmann equation and beam scheme. *International Journal of Modern Physics C*, 9(8):1177–1187, 1998.

538. Yan, G. A Lagrangian lattice Boltzmann method for Euler equations. *Acta Mechanica Sinica (English Series)*, 14(2):186–192, 1998.

539. Yan, G. Recovery of the solitons using a lattice Boltzmann model. *Chinese Physics Letters*, 16(2):109–110, 1999.

540. Yan, G., Y. Chen and S. Hu. A lattice Boltzmann method for KdV equation. *Acta Mechanica Sinica (English Series)*, 14(1):18–26, 1998.

541. Yan, G., Y. Chen, and S. Hu. Simple lattice Boltzmann model for simulating flows with shock wave. *Physical Review E*, 59(1):454–459, 1999.

542. Yepez, J. Lattice-gas quantum computation. *International Journal of Modern Physics C*, 9(8):1587–1596, 1998.

543. Yu, H.-D. and K.-H. Zhao. A new lattice Boltzmann model for a two-phase fluid. *Chinese Physics Letters*, 16(4):271–272, 1999.

544. Zanetti, G. Hydrodynamics of Lattice Gas Automata. *Phys. Rev.*, A 40:1539–1548, 1989.

545. Zhifang Lin; Haiping Fang and Ruibao Tao. Improved lattice Boltzmann model for incompressible two-dimensional steady flows. *Physical Review E*, 54(6):6323–6330, 1996.

546. Ziegler, D.P. Boundary conditions for lattice Boltzmann simulations. *J. Stat. Phys.*, 71(5/6):1171–1177, 1993.

547. Zienkiewicz, O.C. *The Finite Element Method in Structural Mechanics*. McGraw-Hill, Maidenhead (UK), 272 pp, 1967.

548. Zienkiewicz, O.C. and R.L. Taylor. *Finite Element Method - Basic Formulation and Linear Problems, Vol. 1*. McGraw-Hill, New York, 1989.

549. Zienkiewicz, O.C. and R.L. Taylor. *Finite Element Method - Solid and Fluid Mechanics: Dynamics and Nonlinearity, Vol. 2*. McGraw-Hill, New York, 1991.

550. Zorzenon Dos Santos, R.M. Using cellular automata to learn about the immune system. *International Journal of Modern Physics C*, 9(6):793–799, 1998.

551. Zou, Q., S. Hou and G.D. Doolen. Analytical solutions of the lattice Boltzmann BGK model. *J. Stat. Phys.*, 81(1/2):319–334, 1995.

552. Zou, Q., S. Hou, S. Chen and G.D. Doolen. An improved incompressible lattice Boltzmann model for time-independent flow. *J. Stat. Phys.*, 81(1/2):35–48, 1995.

553. Zuse, K. *Rechnender Raum*. Vieweg, Braunschweig, 1969.

554. Zuse, K. *Calculating Space*. Technical Report Tech. Transl. AZT-70-164-GEMIT, MIT Project MAC, 1970.

555. Zuse, K. The computing universe. *Int. J. Th. Phys.*, 21(6/7):589–600, 1982.

Printing and Binding: WB-Druck, Rieden/Allgäu